Tracking the Weretiger

TRACKING THE WERETIGER

Supernatural Man-Eaters of India, China and Southeast Asia

Patrick Newman

McFarland & Company, Inc., Publishers
Jefferson, North Carolina, and London

LIBRARY OF CONGRESS CATALOGUING-IN-PUBLICATION DATA

Newman, Patrick, 1961–
　　Tracking the weretiger : supernatural man-eaters of India, China and Southeast Asia / Patrick Newman.
　　　　p.　　cm.
　　Includes bibliographical references and index.

　　ISBN 978-0-7864-7218-5
　　softcover : acid free paper ∞

　　1. Tigers—Asia—Folklore.　2. Leopards—Asia—Folklore. I. Title.
GR265.N49　2012
398.36975'54—dc23   2012029671

BRITISH LIBRARY CATALOGUING DATA ARE AVAILABLE

© 2012 Patrick Newman. All rights reserved

No part of this book may be reproduced or transmitted in any form or by any means, electronic or mechanical, including photocopying or recording, or by any information storage and retrieval system, without permission in writing from the publisher.

On the cover: Print of man-eating tiger shot by Indian hunters, 1911.

Manufactured in the United States of America

McFarland & Company, Inc., Publishers
　Box 611, Jefferson, North Carolina 28640
　　www.mcfarlandpub.com

For Jessie, Jimmy, and Lucy

Acknowledgments

I am indebted to the following: above all, Anna, without whose support this book could never have been written; Paul T. Barber, of the Fowler Museum of Cultural History at the University of California, Los Angeles, who read an early version of the manuscript and made many valuable suggestions, some of which prevented me making a complete fool of myself (though all remaining mistakes are, of course, my own); Gill Jackson, who gave astute advice on an even earlier version of the manuscript; Jet Bakels, of the Royal Netherlands Institute of Southeast Asian and Caribbean Studies in Leiden, for permission to quote from her work; Roland Smithies, of Luped Media Research, for his expert picture research; the staff of the British Library, and the staff of the Special Collections Centre at the University of Aberdeen Library, for their patient help; Belinda and Emma, for their constant encouragement; Stewart Ross, for inspiring me all those years ago; and Kevin, Mark, and Moo, for everything.

Table of Contents

Acknowledgments	vi
Preface	1
Introduction: The World of the Weretiger	3
1 A Colonial Menace: Jungle-Wallahs and Man-Eaters	9
2 People and Tigers: Together in Life and Death	27
3 Killers and Killed: Propitiation and Appeasement	55
4 Fighting Back: Jungle-Wallahs to the Rescue	76
5 Shapeshifters All: Like Weretiger, Like Werewolf	93
6 Tigermen in Malaya: Negritos and Jambi Men Accused	113
7 Beast People: Weretigers and Wereleopards in India	135
8 Beast Master: The White Sadhu and the Ultimate Terror	167
Chapter Notes	185
References	195
Glossary	200
Index	201

Preface

This book considers the superstition and folklore attached to man-eating tigers and leopards by the people of Asia on whom the animals mainly preyed in the colonial era. It is the result of many months spent researching colonial memoirs and studies by folklorists and anthropologists, and arises from a lifelong fascination with both man-eaters and European colonial history, especially the history of the British in Asia. And being British myself, I make no apology for writing mainly, though by no means exclusively, about British India and Malaya.

The decline of the tiger in the wild is well known. Around 1900, perhaps 100,000 tigers still occupied most of the Indian subcontinent, Burma, Siam (now Thailand), the Malay Peninsula, French Indo-China (Laos, Cambodia, and Vietnam), the Dutch East Indies (Sumatra and Java), Bali, Korea, and much of China, eastern Russia, and Central Asia. Due to habitat loss, hunting and, most recently, poaching, only around 3,000 are now thought to survive. The Balinese subspecies became extinct in the 1930s, and the Javan and Caspian subspecies were last seen in the 1970s.

Side by side with the decline of the tiger has been the loss of the traditional ways of life of tribespeople whose age-old beliefs are at the heart of so much tiger lore. In the 1970s the Batek Negrito hunter-gatherers of the Malay Peninsula told American anthropologist Kirk Endicott that the spirits would destroy the world if the Batek ever left the forest[1]; today there are only a few hundred Batek left, clinging to their traditions against all odds. Similarly, in the 1930s the Mlabri hunter-gatherers of northern Thailand told Austrian anthropologist Hugo Bernatzik that the spirits would send tigers to kill them if they ever gave up their wandering ways[2]; again, today there are only a few hundred Mlabri left, many of whom have been forced to settle down.

Tigers may be on the way out, and many tribespeople in Asia, as elsewhere, face an uncertain future at best, and a bleak one at worst, but tiger lore takes longer to die, often adapting to accommodate the disappearance of tigers. In Java there was once reputed to be a village called Lamongan inhabited solely by hereditary weretigers, or "tigermen." By the 1950s, reported Dutch big-game hunter J. C. Hazewinkel, people were accounting for *their* "disappearance" by saying they had died out, their marriages having become childless.[3] American anthropologist Robert Wessing has observed that not only does the tiger retain an important symbolic function in Javanese people's lives but, poignantly, many Javanese refuse to accept it has gone from their island, remaining convinced it still roams the forest.[4]

As a Malay proverb has it, the tiger dies, but his stripes remain.

Introduction: The World of the Weretiger

The tiger's stripes are on the outside: man's are on the inside.
— Bhutanese proverb

Look in almost any second-hand bookstore and somewhere there you will find an old memoir about hunting tigers in the distant days of British India. Just to open one of these faded tomes, with their grainy black-and-white trophy shots, is to enter a world in which night after night, it seems, roaring man-eaters prowled around isolated villages while the helpless inhabitants cowered inside their huts, praying for a fearless white savior to end their tormentors' long reigns of terror.

Whatever the truth of the matter, when this world is skillfully evoked it is impossible not to be captivated by accounts of sitting alone through the long dark hours in a precarious tree *machan*, or platform hide, waiting silent and still for a man-eater to creep back to the bloody remains of its latest victim sprawled out on the ground below; the hunter knowing all the while that the unseen killer might well be stalking *them*.

But more fascinating still, to my mind, are those accounts that record the colorful folklore surrounding man-eating tigers and leopards alike: that the killers were the agents, even manifestations, of enraged or malevolent deities (or spirits or demons); that they were urged on by the thumb-sized spirits, or ghosts, of their victims riding shotgun on their heads; and that some of the worst of them were vengeful or bloodthirsty black magicians who somehow shapeshifted into killer big cats (Figure 1).

Such stories are not unique to British India. Wherever tigers have roamed you find similarly intriguing tales, with French Indo-China, and the Malay world of the Dutch East Indies and the British-run Malay Peninsula, offering particularly rich pickings.

Figure 1. Enigmatic American illustrator Mahlon Blaine's depiction of a tiger, said to be a transformed Negrito, tormenting a Malay girl in her home in Pahang, sometime in the early 1860s, before killing her and drinking her blood, as recorded by Hugh Clifford. (From the 1927 edition of *The Further Side of Silence* [London: William Heinemann]. The back cover blurb stated — quite fictitiously — that "Mahlon Blaine has ... nosed into many a Malay river for strange cargo and shipped many a Malay crew. He thinks that Sir Hugh Clifford has an uncanny knowledge of native psychology and can substantiate many of the stories by his own experiences.")

All around the world, people untouched by the cold hand of reason have historically sought such explanations for unprovoked attacks by wild animals. Indeed, people worldwide have traditionally sought supernatural explanations for all misfortunes great and small, from earthquakes to earache, that they cannot otherwise explain, and against which they feel powerless. And the poorer they are, the more helpless they feel, and the more vulnerable they are to exploitation of their fears by priests or shamans. If ever there was a useful check on antisocial behavior it was saying that the gods punish wrongdoers by sending man-eaters to kill and eat them, and so need regular propitiation.

Human nature being what it is, all around the world, too, outsiders and other minorities have long been persecuted in times of crisis for supposed witchcraft, not the least of which was shapeshifting, resulting in the same motifs worldwide. Thus an account may be typically Indian in that it incriminates a Hindu ascetic, or sadhu, who, as a devotee of the mighty Shiva, the fierce destroyer (and, as Pashupati, Lord of the Beasts), casts himself in his image and, suspiciously enough, even sleeps like him on a tiger's skin, yet at the same time it may contain such universal motifs as a posse of vigilante villagers following a trail of paw prints that conveniently lead straight to the accused's dwelling.

All too often, though, it was just such marginalized people as sadhus who were most in danger of falling victim to a man-eater. At the same time, despite the risk of reprisal, at least some people, it seems, driven to desperate measures by poverty, exploited and even cultivated their reputations as shapeshifters for personal gain, threatening gullible and frightened people with transforming if they did not give them alms; an especially effective ruse when a man-eater was on the loose.

As no-nonsense servants of empire, most colonials dismissed all talk of supernatural man-eaters as the hysterical reaction of simple-minded savages. Convinced of their enlightened and civilized superiority, they likened the countries under their control — especially the remotest, most "backward" areas, where man-eaters were most prevalent — to medieval Europe, seeing them as being held down by the grip of foolish superstition. And when it came to lynch mobs and exploitative priests they were right. But it is easy to scoff at "native superstition" from the back of an elephant with a powerful rifle in your hand, and are not compelled by poverty to enter the forest on foot every day, alone and unarmed, to gather firewood. And given the terror man-eaters spread it is hardly surprising many people should think them in some way supernatural.

Introduction

There were superstitions enough about tigers and their natural prey. In 1919, Captain Henri Baudesson, a railroad surveyor and missionary in French Indo-China, noted that the "Moi," or "savages" there — the upland tribespeople now more commonly called the Muong — said a tiger abandons its prey uneaten if it tears one of the animal's ears. In 1912, ethnographer Edgar Thurston wrote that the Muduvar and other tribespeople of southern India held that tigers and leopards alike abandon their prey if it happens to lie aligned north to south. And in 1957, sportsman and retired Indian Army officer Arthur Powell reported that professional *shikaris*, or native hunters, throughout India believed tigers and leopards abandon their prey if it chances to lie on its left side.[1]

What the reasons for these beliefs were I have no idea.

Historically, villagers in Asia have known from centuries of living alongside them that tigers and leopards normally give people a wide berth. So when one turned man-eater villagers turned to supernatural explanations: especially if, as often happened, that animal, through a combination of luck and increasing natural wariness, evaded all attempts to kill it and continued to prey on people for months or even years on end with what seemed like extraordinary cunning.

Take the following case reported from Burma by sportsman Sydney Christopher in 1916. A notorious man-eating tigress would lie in wait all night watching the shikari sitting up in a machan over the remains of her latest victim, then pounce on him when he climbed down at first light. She was only outwitted when a shikari called Lu Hpay hid inside a hollow tree and shot her through a small hole he had bored in the trunk.[2]

Consider too the following case reported from the Malay Peninsula by Arthur Keyser, winner of the 1882 Australian lawn tennis championships and author of the first novel set in the peninsula — *An Adopted Wife* (1893), a complicated romance about the capture and torture of a young Englishman by dastardly Chinese tin-miners — and a civil servant there in the 1890s. Night after night a tiger killed a Chinese coolie at the spot where a gang gathered for work after dark, so the local superintendent of police, a man called Syers, decided to disguise himself as a coolie and wait there with his gun each evening. The local Malays laughed, saying that as the tiger was so obviously supernaturally intelligent it would never fall for the ruse. After several nights with no sign of the tiger the sleep-deprived Syers had to concede defeat. The next night the tiger took a coolie from the same spot.[3]

More commonly, long-established man-eaters traveled up to twenty miles between kills while patrolling their extensive territories along habitual beats. Colonials typically ascribed this to a combination of natural cunning and cowardice. Forest officer Frederick Hicks even reckoned it was clear evidence of a guilty conscience, pure and simple.[4] But to local villagers, such behavior—which tended to keep a man-eater one step ahead of its pursuers—was equally clear evidence of supernatural intelligence.

Many man-eaters became so wary of their pursuers that they would abandon a kill at the slightest disturbance, never to return, which meant they frequently ate little or nothing of their victims, leading them to kill much more often than they would otherwise have needed to: sometimes, as with Keyser's tiger, on an almost daily basis, and occasionally even more frequently. This greatly added to the supernatural reputations of man-eaters.

So did the way big cats kill their prey, which is to suffocate it by clamping their jaws around its windpipe. This led to the mistaken belief among colonials and locals alike that big cats suck their prey's blood before eating it.[5] When man-eaters repeatedly abandoned victims uneaten, therefore, people saw the deep puncture marks made by their prominent canines in the victims' throats, and concluded that the animals sought out victim after victim not so much from a hunger for their flesh as from an almost insatiable desire to drink deep of their blood.

Also, if as sometimes happened a long-established man-eater *was* eventually driven out of its territory by persistent attempts to kill it, then its sudden disappearance could cement its supernatural reputation in its former territory, while its sudden appearance in a new territory could gain it instant supernatural accreditation there.

Then there is the fact that a man-eater usually attacks from behind. Colonel Arthur Locke, Kemaman District Officer in the east coast state of Terengganu in the Malay Peninsula in the early 1950s, once tried to tell some Malays that it did this to approach its victim unseen, but they rejected the idea out of hand. The real reason, they said, was that on each person's forehead is inscribed a verse from the Koran declaring humans to be superior to all other creatures: an inscription no tiger could face. In other words, if confronted by a tiger you must have the courage to stand and face it, so it can see the inscription. Victorian sportsman Brigadier-General Reginald Burton too wrote of Indian man-eaters that one of their most remarkable features was a reluctance to face man, that they would often retreat if boldly faced. He also recalled a tigress that, having left a corpse sitting up against a bank, dared not return and face its staring eyes. Most victims, however, are found face down. Again, Malays said this was because a man-eater dared not look its victim in the face. In fact, most victims are found face down because they fall on their faces when attacked from behind, and remain that way during the drag.[6]

Finally, victims are often found with their head, hands, and feet untouched. This could be interpreted as compounding evidence of the tiger's guilt and shame: not only could it not bear to look its victim in the eye, it dared not eat their most obviously human features. In fact, if undisturbed, a man-eater eats every last scrap, including the blood-soaked clothing. But it usually starts by rasping off the skin from heels to neck with its rough tongue, then eats the choice, tender buttocks and other soft parts. The bony hands, feet, and head it eats last, so if it is disturbed into abandoning its meal, these are the parts most likely to be found untouched; with the corpse lying face down.

Whether man-eaters are said to be shapeshifting black magicians, or vengeful demons guided by their victims' malevolent spirits, then the legacy from Asia in the colonial era, I hope you will agree, is a collection of quite marvelous stories from a long-vanished and now all but forgotten world.

1

A Colonial Menace: Jungle-Wallahs and Man-Eaters

> *"The jungles seemed hushed in expectation—no morning hum of insect life. All Nature seemed to be holding her breath to see the terror of the jungles die. Then from far ahead we heard the man-eater on his prey: that never-to-be-forgotten sound—the gnawing of human bones."*—Leonard Handley, *Hunter's Moon*, 1933, 157–8

Man-eaters were a fact of life wherever people encroached on the tiger's domain in India in the days of the Raj, and not least in the then still heavily forested central highlands: the Satpuras, the long chain of hills south of the now dam-threatened Narmada River, which runs almost 800 miles from its source at Amarkantak to the Gulf of Cambay, north of Mumbai (formerly Bombay), and is Hindu India's second holiest river, after the Ganges.

By the 1850s the British controlled a large area of central India—now, by and large, Madhya Pradesh—but not until Crown rule replaced East India Company administration following the 1857–1858 mutiny did they really begin to exploit its resources, when much of it was constituted as the Central Provinces (CPs) and earmarked for "uplifting."[1]

In the wake of the revolt the new government forged ahead with building a network of railroads right across India. This required millions of ties, not to mention a constant supply of firewood to stoke the engine furnaces. Yet so undeveloped was British forestry in India that ties initially had to be imported from England, Norway, and Australia.[2]

Vast as the largely teak and sal forests of the CPs still were, they had been shrinking steadily since at least the sixteenth century, cut down by waves of Aryan Indian immigrants for firewood, lumber, and farmland, and disappearing altogether from many river valleys and other lowland areas. This immigration had petered out by the 1850s, but the new government had no intention of allowing the remaining forests to be destroyed by unregulated commercial felling, or by what it saw as the wasteful and destructive practices of the various tribespeople who lived in them, the majority of whom were Gonds.[3]

Increasingly the Gonds were settling down to farm alongside their Aryan Indian neighbors, but many of them, together with other hill tribes such as the Korkus, Kols, Baigas, and Bhils, still lived a semi-nomadic existence and practiced slash-and-burn shifting cultivation: clearing and burning a patch of forest then sowing it once or twice before repeating the process elsewhere. It is easy to say now that this allowed full forest regeneration in the long term, but back then the British saw only that it resulted in scrub; pioneering CPs forester James Forsyth declaring that "a second growth of timber on such land can never be expected if left to nature." The worst offenders—"the most terrible enemy to the forests we have anywhere in these hills," as Forsyth put it—were reckoned to be the Baigas, keen hunters also known as Bhumias, or "Lords or Sons of the Soil," who resolutely spurned the plow, believing it defiled the breasts of Mother Earth.[4]

The British therefore hurried to exploit the forests of the CPs and elsewhere in India, setting up a forest department and sending in newly recruited British forest officers, with teams of equally green Indian foresters, rangers, and forest guards, to demarcate the most valuable lumber areas as Government Reserved Forests, and to hire tribesmen — Gonds, in the main, in the CPs— as cheap coolie labor to extract the lumber: after first ringing such areas with masonry boundary pillars, to scare away other tribespeople.[5]

For the tribespeople it was the beginning of the end of their old way of life, though they were used to being exploited. With or without the connivance of their white masters, Indian servants and officials had long routinely helped themselves to whatever they wanted from tribal villages under a custom called *begar*, or the purveyance system. In 1844 William "Thuggee" Sleeman noted that even when a sahib ordered his servants to pay for supplies, the servants simply pocketed the money then, to prevent the villagers going to the sahib, resorted to such ruses as saying he was a tiger and would eat them.[6] This was a threat the servants knew the villagers would take seriously.

Begar persisted well into the twentieth century. Verrier Elwin, pastoral officer at the multi-faith Society for the Service of the Gonds ashram, or retreat, at Karanjia, near Amarkantak, in the Mandla district, in the

1930s — and later a noted if somewhat controversial anthropologist and tribal activist — wrote that people spoke almost in awe of a former range officer who always paid in full for all he took, and that the Baigas still treasured the memory of an Englishman who paid them their full wages.[7]

For the Baigas in particular the 1860s saw the end of their Sukhiraj, or "Golden Age," and the start of their Kali Yug, or "Age of Darkness." Elwin noted that when the first train came into Mandla a Baiga sacrificed a cockerel before it, propitiating it as an evil spirit. Ironically, Baigas themselves were renowned for their honesty. In 1906 Colonel Alfred Bloomfield, retired Indian Army officer and former deputy commissioner of the Balaghat district, recalled often hearing Baiga witnesses in his court exclaim, "What! — a Baiga tell a lie? Never!"[8]

Captain James Forsyth was just the kind of young man the British needed in the early days of the CPs. Born in 1838 in the tiny parish of Morham in Haddingtonshire (now East Lothian), the second son of the Rev. James Forsyth, from an early age he was shooting-mad and dead-set on a life of *shikar*, or hunting, so after graduating with honors from Marischal College in Aberdeen he secured an appointment with the Indian military. Aged nineteen he arrived in India just in time for the Mutiny, following which, wrote his father, "duties of the most revolting nature fell to him ... such as superintending the execution of those who were caught red-handed in the atrocities perpetrated by the rebels, and who were condemned sometimes by scores to be blown into fragments from the guns." Then in 1861 he was garrisoned with the 25th Punjabi regiment in Jabalpur, in the district of that name, at the base of the Maikala Range near the head of the Narmada valley.[9]

Only five years later Jabalpur would be a main stop on the Great Indian Peninsular Railway, a bustling civil as well as military hill station, and the plum posting in the CPs, but when the young Scot received his orders to march there no one could even tell him where it was, despite it being where in the 1820s Sleeman began his campaign against the Thugs, the supposedly India-wide secret society of murderers who strangled strangers in the name of Shiva's consort Kali, dread goddess of death.[10]

On arrival in Jabalpur, Forsyth was astonished by the beauty of its setting, where white marble cliffs rise up on either side of the Narmada, and settled down to enjoy himself. He had always admired the gallantry of his Sikh comrades, and for the next year reveled in their company at frequent parties in which, he recalled, "plenty of champagne, brass bands and songs that were sometimes very much the reverse of hymns, bore the most prominent part." He also spent a month's leave indulging his passion for shikar in the surrounding hills.[11]

But overall he found peacetime soldiering dull, so he was delighted in early 1862 to be chosen as one of a handful of army officers to be sent out as "conservators" to explore and assess the vast forests of the CPs. His first task in his new role was to proceed to the Hoshangabad district and establish a permanent forestry base at the Korku village of Pachmarhi there, on a park-like plateau of the same name, riddled with gorges and waterfalls, in the cool heights of the Mahadeo Hills. This was only fitting, for it was he himself who had "discovered" Pachmarhi for the British back in 1857, when leading his troops along an old caravan route.[12]

In years to come forest officers—or jungle-wallahs, as they were known—would travel in typical British style, with trains of servants and baggage animals carrying sundry large tents and every conceivable item to furnish them with, including carpets. But in the then largely unknown ranges of the CPs—"where all is chaos to the unguided traveller; where hill after hill of the same wild and undefined character are piled together; where the streams appear to run in all directions at once," as he described them—pioneering forester Forsyth sensibly opted to travel light. A few ponies carried a small tent for his handful of Indian servants and foresters, and a larger tent for storing building materials at Pachmarhi, while a camel carried all his own gear: a one-man tent, a folding bamboo charpoy, or bed, a cane stool, a small folding table, a portmanteau, or trunk, his precious guns, a brass basin and stand, and what he called "a good supply of such eatables and drinkables as are not to be had in the wilderness." The latter included some Scotch whiskey from back home, and several bottles of fine claret, while the camel in question was Junglee, "a camel among camels" he had caught running wild when hunting mutineers and had thereafter served him well without ever really being tamed. "Poor Junglee!" he recalled. "He afterwards ended his days under the paw of a tiger."[13]

On Forsyth's recommendation Pachmarhi would later become a sanatorium for troops and a thriving hill station, but the young Scot would be its first British inhabitant. On arrival he set about hiring local Gonds and Korkus to build a stone bungalow, which he would later name Bison Lodge. Initially this proved difficult, for they were shy and suspicious. The Korkus in particular were reluctant to help. "With some truth," wrote Forsyth, "they feared we were come to break up their much-loved seclusion, and untrammelled barbarism."[14]

The Korkus also feared losing the revenue they extracted every February from the thousands of pilgrims who trekked from far and wide up to Pachmarhi to a shrine to the Mahadeo, or "Great God," himself—Lord Shiva—where a stream flows from a cave in a cliff. The terrified pilgrims associated the tribespeople with the myriad demons, diseases, and beasts they believed guarded the way, and readily paid up to safeguard their skins. In times past, though, many had not been afraid to hurl themselves off the cliff in the ultimate offering to Kali and Kala Bhairava—a manifestation of Bhairava the wrathful, the tiger-toothed embodiment of fear and the most terrifying of Shiva's various incarnations—until the British put a stop to such things in the early nineteenth century. (Another shrine to Shiva, called Omkar, in the Nimar district, on Mandhatta, an island in the Narmada, was similarly guarded by Bhils. Pilgrims once flung themselves off that, too.)[15]

To win the Gonds and Korkus over and persuade them to help build Bison Lodge, Forsyth gave them flour—and gave three or four local Gond *thakurs*, or chiefs, gunpowder and bullets—and organized a hunt with hundreds of Gond and Korku beaters and some twenty matchlock-wielding Gonds, the result of which was a fine mixed bag of bison, sambar (a large deer), and miscellaneous smaller game. The full-blown celebrations that followed both amazed and disgusted the young man who had himself partied so hard at Jabalpur, as they wolfed down half-cooked meat and entrails, guzzled spirit fermented and distilled from the fleshy flowers of the mahua tree (*Madhuca longifolia*), and sang and danced long into the night. (Forsyth himself sampled mahua spirit on at least one occasion, his expert judgment being that when well made and mellowed by age it was "by no means of despicable quality, resembling in some degree Irish whisky," while Elwin wrote that the Gond vision of Hell was a forest with no mahua trees; and that of Heaven one with no forest guards.)[16]

Despite all the masons and bricklayers who had reluctantly accompanied Forsyth up from the plains absconding, construction proceeded rapidly, if somewhat haphazardly, the young Scot complaining that "Regular industry ... was not to be got from these unreclaimed savages; and there were seldom half of those on the muster-roll actually present. Every now and then, too, they would walk off in a body, and have a big drink somewhere for a couple of days, returning and setting to work the next morning without appearing to think a word of explanation necessary."[17]

Finally, though, the lodge—complete with a newly plowed one-acre garden—was ready, and Forsyth was able to set out and begin exploring and assessing. Meanwhile, the fledgling Forest Department began receiving young men from Britain, and towards the end of 1866 the seasoned Scot briefly shared Bison Lodge with a fresh-faced Englishman, Frederick Hicks. To Hicks's delight, Forsyth immediately recruited his help in research for the second edition of his 1863 book *The Sporting Rifle and Its Projectiles*: experiments with the already famous "Forsyth's shell," an explosive bullet for shooting tigers and other big game.[18]

Hicks too was a clergyman's son, being raised on the banks of the Suffolk Stour in the Essex village of Sturmer where his father, William, was Rector of St. Mary's. There he enjoyed a carefree upbringing with his older brother Herbert, the two boys spending much of their time fishing, shooting, and ferreting — largely in the capacity of poachers— and riding. Herbert followed his father into the Church, while Frederick crammed for the Home Civil Service entrance exam. But by his own admission, such was not his forte, and he duly failed. His father had been in the navy, being wounded by a cannonball as a young midshipman on HMS *Conqueror* at Trafalgar in 1805, so when the British government offered land in Canada to sons of officers who had fought in the Napoleonic Wars, young Frederick's education took on a more practical bent, in the form of carpentry, shoeing horses, farming, and blacksmith's work. But his gaze soon turned away from Canada. "The glamour of the East had got hold of me," he explained. "I had been reading a lot about the glories of big-game shooting that was to be had there, and I also made the acquaintance of a man from India who still further inflamed my brain with his accounts of the doings in these eastern lands." So his father secured him an appointment as a forest officer in India instead, and in 1866, a bachelor in his twenties, he set sail for Bombay.[19]

A brief, abortive attempt at farming in Australia in 1882 aside (after he was temporarily invalided out of the service following a pig-sticking accident), Hicks was a forest officer in India for the next thirty-two years: seven of them in the southern princely state of Mysore in the 1870s, when he married and began a family, and the rest in the CPs. For many years he could not afford to visit England. Then in 1887, after twenty-one years away, he finally went home on three months leave to put his sickly son in the care of cousins and find him a boarding school. It was a salutary experience, for he discovered he had lost touch with his past and no longer belonged. He never returned, staying on in India after he retired in 1898, by which time his son had joined the Indian Police. (For his son to have followed in his footsteps, Hicks would have had to pay for him to train at

1. A Colonial Menace

the newly established Forestry School at Cooper's Hill in Surrey, something that, having no private means, he could not afford. And having no private means, he could not have afforded the lifestyle he had grown accustomed to in India — one based around having servants— in retirement in England anyway.) But he never forgot the village of his youth. "Dear old Sturmer —" he wrote, "how often in this land of heat, dust and scorching winds, have my thoughts flown to your cool green banks, overhanging willows and deep clear pools."[20]

Hicks was gripped by the sap-scented forests of India from the moment he arrived at Bison Lodge to begin his first three-year posting. "What delightful memories of my early forest life are called up by those beautiful evergreen covered ranges," he enthused, "for it was among them that I first opened my forest career in the prime of my youth, hope, strength and confidence, with all the rosy promises of life before me." Engaged mainly in railroad tie production, straight away he found himself in charge of 500 coolies and a team of elephants, animals bought two years earlier by Forsyth at the great annual fair at Sonpur on the Ganges. He quickly took to his new life. With an elephant pulling the plow, at Bison Lodge he dug a large plot to grow English vegetables in, while out in the forest he once fined two coolie brothers for slacking, and the next day the aggrieved pair attacked him, one wielding an ax. An accomplished amateur wrestler and boxer, Hicks flung the two much smaller men to the ground and knocked them both out. When they came to he gave them the choice of being tried for attempted murder, or being tied to a tree and receiving twenty lashes from his orderly, Dilliput, a hefty Afghan Pathan, or Pashtun. They prudently chose the latter, and thereafter gave Hicks no more trouble. He later even promoted one of them for good work.[21]

After learning the ropes in Hoshangabad, in 1869 Hicks was posted to the immense, marshy Ahiri Forests of lower-lying Chanda where, while moving from camp to camp, he did not see another colonial or hear English spoken for months at a time. As he recalled: "I was practically my own master, as long as I satisfied my superiors with a report once in a while to the effect that my work was progressing satisfactorily; for which, moreover, they had to take my word, for no one would come out to such an outlandish place."[22]

Day in, day out, Hicks, his rangers and foresters, and forty coolies beat their way through the undergrowth of selected forest blocks, recording as they went the number, size, species, and condition of all trees over a certain girth. Having checked a block they hacked a demarcation line around it. It was back-breaking work in the stifling heat, and entangled with the spear-grass was a mass of "cow-itch," or cowage: a climber with pods that at the slightest touch release a shower of minute spines, which work their way into the skin and itch maddeningly. They also had to keep an eye out for hornets—"big black brutes some four inches in length," Hicks reckoned, "with a sting like a brad-awl, three of whom have been known to kill a man"—and once when they disturbed a nest they had to torch the undergrowth to save themselves from the swarm.[23]

Forsyth had similar troubles. In Mandla his party was once attacked by a swarm of bees and one baggage pony, unable to shake off its load and flee with the others, was stung to death on the spot. Another time he disturbed a nest of red ants in an overhead tree when hot on the trail of a tiger on elephant-back, and within moments himself, the elephant, his mahout, or driver, and the mahout's boy (who sat behind the howdah, or seat, with a stick, ready to steer the animal in a charge) were all being bitten mercilessly. All thought of the tiger was forgotten in the rush to plunge into the nearest river, but even then each ant left its head and pincers clamped in place, and it took half an hour to pick them all out.[24]

But there were much more serious threats to the health of early forest officers, notably "jungle fever," or malaria, which in those days was still believed to be caused by a miasma, or noxious vapor, rising out of rotting vegetation, especially in swamps and marshes—"malaria" literally means "bad air"—the link with mosquitoes not being demonstrated until the late 1890s.[25]

Hicks suffered from malaria for the first fifteen years of his career, being so feverish at one point in Chanda that a fellow colonial who chanced by his camp found him delirious on his charpoy with a farewell letter to his father tucked under his pillow. He also suffered from dysentery: "To boil my water, or to look after my kitchen," he admitted, "were matters that never entered my head, or if they did, I probably scoffed at such molly-coddling ideas." In Mysore he also caught cholera, but he was newly married by then, and thanks to his wife he recovered; and from then on, with her to look after him, he stayed more or less healthy.[26]

Disease was an ever present danger to a forest officer's servants, staff, and coolies, too. In 1894 Dilliput contracted cholera and died. "The poor fellow had been my faithful servant for many years," wrote Hicks, "and I felt his loss sadly. His last request was for a drink of tea. I saw that there was no possible hope for him, so I complied ... and having drunk it, he suddenly staggered to his feet and saluted me, and immediately fell back dead."[27]

1. A Colonial Menace

Bachelor Forsyth was not as lucky as Hicks, either. He contracted malaria early on in India and suffered from it continually. (One scorching day, when confined to his tent by fever, it became so hot and stuffy inside that finally he staggered out and lay "like a tiger" in the shade of some riverside evergreen bushes.) In 1870, by which time he was assistant commissioner of Nimar, he returned to Scotland on extended leave to care for his father, who had himself taken ill, and spent the winter writing *The Highlands of Central India*. His father recovered, but Forsyth died while on a visit to his London publisher in May 1871, aged thirty-three, just before his seminal book's publication, and was buried in Kensal Green Cemetery, in an unmarked grave. As he himself wrote of pioneering foresters in India, "there is not one of them whose health did not, after a few years, give way under the combined assaults of malaria and a fiery sun."[28]

The job may have been hard and hazardous, but a forest officer played hard too. Invariably he spent much of his spare time shooting: that and adventure in general usually being what drew him to the job in the first place. Indeed, he was encouraged to shoot in order to keep himself occupied, get to know his district, and toughen him up. Hicks once heard Sir Richard Temple, CPs chief commissioner from 1864 to 1867, declare any forest officer not fond of shooting "not worth his salt."[29]

The best time to hunt tigers in particular was in the hot months at the end of the dry season, before the monsoon rains, when game and cattle alike were concentrated around the few pools left in the lower reaches of largely dried up rivers. Hicks recalled the satisfaction of returning to camp at sundown after a hard day's shooting for a welcome peg of whiskey, a hot bath, and dressing for dinner — over which he would enthrall his family with an account of the day's battles — then adjourning with a cigar to the side of a roaring fire while servants skinned the tiger he had bagged. "What more delightful and healthy life could any man wish for?" he asked.[30]

The tables were turned, though, one day in January 1890, in Mandla. A tigress he had wounded and was following up on foot charged back with a roar, knocked him to the ground, and sunk her teeth into his left hip then hand: "Then everything seemed to go round and round, and the last thing I remember is the tigress with her head raised in a listening attitude with a far-away look in her eyes, with the tendon of my wrist hanging hooked on to her eye-tooth, jerking it from time to time and sending excruciating shocks of pain up my arm — and then blank!" When he came to, the tigress was slumped dead across his chest.

His men had all bolted, but Hicks managed to stagger the three miles back to camp, where two of them sheepishly came to his aid. Then he saw his wife who, despite a tearful Dilliput telling her the sahib was dead, had grabbed a gun and begun organizing a rescue party. "I at once made my supporters stand off," wrote Hicks, "while I lit a cigar and putting my left hand behind my back, met her jauntily whistling a tune as if nothing had happened, though my boots, by squelching with blood, gave the show away rather."

Still in shock, that evening he insisted on dressing for dinner and doing his paperwork with his *moonshee*, or secretary, as normal. But as usual with such a mauling his wounds were already infected and his temperature was soaring. Knowing he was in for a bad time and might not recover — there were no antibiotics then — he gathered his staff and, with his maps spread out before him, gave them instructions for the rest of the dry season, asking each in turn to swear to do his duty and not take advantage of his absence in any way. This they did, he wrote, with tears in their eyes: "That night, and for many weeks afterwards, I was in high delirium and completely off my head, a special servant having to be kept in order to hunt away imaginary tigers from the various corners of my tent…. For six months I lay thus in the intensest of agonies, hovering between life and death. I prayed for death to release me from my pain. I begged those around me to let me die."

But his wife would have none of it, and nursed him back to health. She also took over many of his duties, giving his staff their orders, checking the accounts, and even writing progress reports. As for his staff, he wrote, "let who will say what they like of the native — every man among them kept his word to me and worked like a brick." It was nine months before Hicks was well enough to resume work. He at once took to shooting again, he noted.[31]

Hicks's coolies would probably have attributed his recovery to the death of the tigress, for among the tribespeople of central India, at least, it was said you would only survive a mauling from a tiger or leopard if the animal was killed and its body brought before you as proof of its death. Hicks himself recalled a mauled man refusing to go to hospital for just this reason. Hicks looked long and hard but failed to find the tiger, and sure enough the man died. Then again, Forsyth recalled his faithful personal shikari, "the Lalla," being mauled in 1866 by a tigress he had wounded and was following up on foot while the young Scot lay ill back in camp. "The fire of delirium was then in his eye," he wrote, "and he raved of the tiger's form passing before him, red and bloody. But he recognised me when I came to him, and conjured me to go out forthwith and bring in her body next day if I wished to see him live." This Forsyth did, despite his fever, but alas, the poor Lalla still died.[32]

Besides hunting tigers for sport, forest officers spent much of each dry season hunting persistent cattle-lifters, and man-eaters. An established man-eater might operate for year after year in a territory of 100 square miles or more. Anyone living and working in this area — and that meant tribespeople, mainly — was in danger of attack, but those most at risk included: herdsmen who grazed their livestock in the forest fringes; young women who daily visited the same to collect sticks for firewood, or grass and leaves for fodder and thatch; woodcutters; forestry coolies; and village *chowkidars*, or night-watchmen. Villagers in the grip of fear of a persistent man-eater barricaded themselves indoors at night, or kept lines of fires burning from dusk to dawn. By day they ventured out only in large groups guarded by shikaris, shouting and beating drums as they went. Despite such measures, in time whole villages would be abandoned.[33]

People traveling on foot through a man-eater's territory were also in great danger, notably: pilgrims journeying to and from remote shrines; Banjaras, a low caste of cattle-herders and traders, often called the Gypsies of India, who traveled with fierce dogs wearing spiked collars for protection against big cats; and post or mail runners.

Mail runners carried a staff festooned with rings or bells, banging it on the ground every few steps to alert villagers to their approach and, in theory, scare off snakes and tigers; and presumably also to ward off evil spirits, that being one of the original functions of bells worldwide. One of Hicks's successors in the CPs was Wellington College educated and Cooper's Hill trained "Jungli" James Best, youngest son of the fifth Baron of Wynford — and therefore most definitely a man of private means— and a forest officer from 1904 until 1925, when he retired to Melplash in his native Dorset to keep bees. Noting that mail runners formed the main diet of a notorious tiger in Balaghat, he wrote, "one can imagine the rascal when feeling hungry listening with impatience for the jingle of the bells announcing the arrival of the post and his dinner!"[34]

In Chanda, runners were supposed to bring Hicks his mail — not to mention some cigars, whiskey, and beer — once a month, but often the runner never made it, drowning or being killed by a tiger on the way. In the Upper Provinces (UPs) district of Naini Tal in the Himalayan foothills in 1864, forest officer Thomas Webber heard that a few years earlier a tigress killed so many runners on the Juli road there that eventually the district commissioner ordered two men to run together. First time out, the tigress pounced on the man carrying the mail, but his companion managed to beat her off then carry both him and the mail bag the rest of the way. "The poor man was badly mauled," wrote Webber, "and the tiger's teeth-marks were through an official envelope delivered next morning to the commissioner." (Almost certainly this was the same man-eater cited by sportsman Joseph Fayrer of the Bengal Medical Service as killing eighty people a year from 1856 to 1858.)[35]

In the CPs, one of the worst areas for man-eating was the sparsely populated, gorge-riven, and almost inaccessible Dindori plateau high in the Maikala Range in Mandla and the Bilaspur district around Amarkantak. Every year thousands of pilgrims trekked up to the shrine at the spring above the Kapildhara falls, where there was a large settlement of sadhus, and to smaller shrines nearby. Pilgrims and sadhus became the staple food of tigers in the area until finally when Best visited it in the early 1900s there was only one old sadhu left, the rest having fled or been eaten. Best urged him to leave, but he refused. When Best returned in 1912, he too had been eaten.[36]

People are easy pickings, but why a tiger becomes a man-eater in the first place has long been a matter of much debate. In colonial India it was generally agreed that some tigers acquired a taste for human flesh when famines and epidemics— and there were plenty of both — provided them with a ready supply of corpses: and that when the supply dried up they turned to killing people to maintain their new diet. Other tigers, it seems, took to man-eating when they grew too old and infirm to catch their wary, fleet-footed natural prey— wild boar and deer, in the main — or suffered some accidental injury that stopped them doing so, such as getting quills embedded in the face or paws when tackling a porcupine. And still other man-eaters were tigers prevented from hunting their usual prey by being incapacitated, even if only temporarily, by gunshot wounds.

As India's human population grew, and farming increasingly encroached on its forest fringes—where wild boar, deer, and tigers were most numerous— tigers increasingly preyed on livestock where pasture, not crops, replaced trees and made natural prey scarce, many becoming persistent cattle-lifters. Almost every circle of villages had its resident shikari, who villagers paid to sit up at night, armed with a (usually ancient) small-gauged, single-barreled matchlock, muzzle-loaded with whatever bits of scrap metal he could find, and guard their crops against wild boar and deer. Village shikaris rarely attempted to tackle man-eaters on their own, but they often took pot-shots at livestock-lifters, with the result, according to Best and others, that they wounded more tigers than they killed.[37]

Forest officers and other sportsmen had more powerful and accurate double-barreled rifles. Even so, experienced and responsible ones like Hicks never shot at a tiger unless they were confident of killing it on the spot, and if they did wound it they always tried to follow it up, despite the risk; as Hicks discovered to his cost. As sportsman James Brown of the 79th Cameron Highlanders put it in 1887: "Any man who goes in for tiger-shooting should make up his mind to consider it a point of honour to follow up, even alone if necessary, any tiger that he may *wound*, and follow him up until he either bags him or loses him, for a wounded tiger often takes heavy toll of human lives when left behind." But this was impossible if it retreated into thick cover, or if the light was failing — something that happens rapidly in the tropics — and the most seasoned hand sometimes had a wounded tiger get away. Forsyth noted that in May 1862 in the Betul district he shot dead six tigers and wounded two others, both of which escaped.[38]

Nor was every colonial experienced and responsible. Against his better judgment, in Jabalpur Hicks agreed to take a friend's son out to bag his first tiger. The young man was not really interested — "his thoughts were more with polo, tennis, parties and ladies," recalled Hicks — but it was the done thing, so he went, shot carelessly, and wounded two tigers. Both got away, and one killed a villager who stumbled across it licking its wounds a few days later. That did not make it a man-eater — several at least partially eaten victims had to be recorded for a tiger to be declared such — but it increased the chances of it becoming one.[39]

Not all man-eaters were old or injured, however, and not all old or injured tigers became man-eaters. It may be an oversimplification, but it seems that as people and tigers came increasingly into contact, some tigers, and particularly tigresses with hungry cubs to feed, simply discovered just what easy pickings people were.[40]

As if man-eating tigers were not bad enough, leopards that turned to man-eating spread their own form of terror, sneaking or forcing their way into homes at night to claim their victims while they slept. No one was safe when one was on the prowl. One night in 1906, Best pitched camp by a village of some 500 inhabitants in an area of the Sonawani Range (between Mandla and Nagpur) where a man-eating leopard was operating.

He had just drifted off to sleep when a scream woke him with a start. In seconds the whole village was yelling in panic, followed by his camp-followers. He grabbed his rifle and ran among the huts, shouting to make himself heard as he tried to find out what was wrong. "Panther!" yelled the headman from behind his door. "Where?" cried Best, looking all around. But no one could say. No one had seen anything. It turned out the villager who screamed had merely had a nightmare about the leopard.[41]

Leopards took to man-eating for much the same reasons as tigers did. Edward Baker, deputy inspector-general of police in Bengal, noted in 1887 how increasingly common leopards were around human habitations in the province, attracted by livestock as their natural prey declined. Leopards were also attracted by people's dogs, dogs widely being said to be their favorite food. "For a meal of dog, they will dare anything," wrote Indian Army officer Edward Durand, who once had a pet terrier snatched off his camp bed when sleeping outside his tent in the hot dry season.[42]

As late as 1906, noted Hicks, the government had to deputize a special forest officer for the sole purpose of ridding Mandla of man-eating tigers. The government saw man-eaters as a menace not just because of the toll they took—Dutch academic Peter Boomgaard cites 1,200 people killed annually by tigers and leopards in late nineteenth-century British India—but because they brought forestry and other work to a halt, the coolies fleeing to a man when one pitched up. Man-eaters also panicked villagers into setting swathes of forest alight to try to scare them away, destroying valuable lumber, and frightened herdsmen into giving up their grazing licenses, meaning more lost revenue. Joshua Carrington Turner, a forest officer for thirty years until 1955, recalled a tiger stopping construction of the rest-house for British officials at Paharpani in Naini Tal. "There was only one decision..." he wrote: "to go to Paharpani and, on arrival there, to stand over [the contractor] Harlal's workers with a gun. This, I hoped, would instil them with a sense of urgency as well as confidence."[43]

Such a nuisance did the British government consider tigers and leopards to be that as well as offering rewards, or bounties, of up to 500 rupees for man-eaters, there were smaller rewards for *any* tiger or leopard bagged. These varied over the years, and from place to place, but generally they seem to have been 50 rupees for a tiger and 25 for a leopard.[44]

Hicks reckoned that very few sportsmen did not claim these smaller rewards, which helped offset the cost of hiring beaters and shikaris, but they were really aimed as inducements to shikaris to try to shoot, trap, or even poison as many tigers and leopards as they could. They were well worth having if you were a shikari, and as a result, wrote Best, more shikaris took pot-shots at tigers and leopards, resulting in more of the animals being wounded. The rewards also encouraged some shikaris to become professional *bagh-maris*, or tiger slayers (*bagh* meaning tiger, or big cat). Burton reckoned the bagh-maris of the Dinajpur district of Bengal, who traveled all over the province hunting tigers, often claimed twice for the same animal by presenting the head to one district officer and the skin to another, while Baker suspected that many of the leopard skins and skulls brought in by bagh-maris looking to claim rewards in Bengal actually came from Nepal, where no rewards were paid.[45]

British policy was the same in the Malay Peninsula where, according to Major Frederick McNair, surveyor-general for the Straits settlements, writing in 1878, the reward for an ordinary tiger was 50 dollars (as against 12 for a crocodile).[46] This was for *any* tiger, dead or alive, the consequences of which could be alarming.

One evening Keyser was having tea on his veranda when six Malays pitched up bearing a tiger in the traditional manner, its feet lashed to a pole: only this tiger was very much alive, having been caught in a trap, one almost certainly of the drop-door variety (Figure 2). Dumping it in front of him, they said not to worry, their district *datok*, or chief, was in no hurry for the reward, saluted, and turned to leave. With the tiger thrashing around, sending the tea table flying, a horrified Keyser pleaded with them to stay, to which they replied, "Oh, we dare not stay, *Tuan* [Lord], we are only poor coolies who know nothing about tigers." Luckily for Keyser the local police barracks were close by and some Sikh officers responded to his yells for help. Their Irish inspector ordered the tiger to be locked up in a cell in the local jail. Keyser had visions of receiving the fellowship of the Zoological Gardens for such a prize, but it went on hunger strike, and a few days later the stench from its rotting, uneaten raw pig and pye-dog rations was so bad that its fellow inmates were in uproar; so the inspector had it shot.[47]

Historically, man-eating tigers were rarer in the Malay Peninsula than in India, though by no means unknown to Malays, who lived mainly in coastal and riverside kampongs, or villages, or to the tribespeople of the dense jungle of the interior. And as many Chinese and Indian coolies discovered to their cost, they became much more prevalent when rubber plantations and open-pit tin mines encroached on the interior in the sec-

Figure 2. Local people in many areas of Southeast Asia in the colonial era set traps for troublesome livestock-lifting tigers. Some were designed to kill the animals, while others, like the wooden drop-door, or "sluice gate," device shown here, on an old postcard from Vietnam — though exactly the same kind were used in the Malay Peninsula — caught them alive. When a tiger entered the trap, attracted by the live bait inside (usually a dog or goat), it triggered the release of the heavy door. The tiger would then either be speared or shot dead while still in the trap, or sold on for exhibiting in a zoo (John Bryan/Fourteenacre Ltd.).

ond half of the nineteenth century. In 1939, J. B. H. Thurston, a rubber-planter in the peninsula for some thirty years, recalled one tiger that was said to have killed more than 100 people, most of them Chinese coolies, causing whole plantations to be deserted, and another man-eater that one day chased a Chinese coolie down the road on his bicycle until the coolie fell off. "Quite incapable of rational behavior," he wrote, "the Chinese scrambled to his knees and prayed to the tiger not to kill him. But the tiger came up to the fallen bicycle and stood over it watching the spinning front wheel and taking no notice of the man whatever." Just then a truck came along and the tiger bounded away.[48]

Fayrer wrote in 1875 that, as with foxes back home, he "certainly would not preserve tigers, and would encourage their destruction, but by hunting, rather than by poison or the snare. Of course, when no sportsmen are at hand, they should be destroyed without law, when and where they could be found." But many sportsmen, like Forsyth and famed elephant-catcher George Sanderson, disagreed, recognizing the role tigers played in keeping crop-raiders in check, and thought only man-eaters and the worst cattle-lifters needed ruthless culling. Much of the clamor for exterminating tigers came from back in England, reckoned Sanderson, and not from those in the know on the ground.[49]

Indeed, the image of the tiger as the diabolical enemy of empire was reinforced back in England in boys' wildlife and adventure books, which were full of tales of derring-do by intrepid sportsmen, often with an artist's impression of the hero, with obligatory handlebar mustache and sola topi, or pith helmet, coolly taking aim even as his snarling foe leapt towards him with outstretched claws, while all around him terrified natives fled for their miserable lives. In 1912, prolific hunting, shooting, and fishing author Frederick Aflalo declared in one such book: "The lion, tiger, and leopard ... do not hesitate to attack natives and annually destroy immense quantities of cattle, sheep, goats, and poultry.... Europeans are morally responsible for the safety and well-being of those whose birthright they administer, and they should consider themselves bound to shoot every lion or tiger they may come across, even at some personal risk and discomfort."[50]

In 1910, Sainthill Eardley-Wilmot — who started as a forest officer in 1873 and rose to become inspector-general of forests, no less — reflected more soberly on the situation: "The tiger is supposed to be vermin even when he lives twenty miles from the abode of man, and is quite incapable of harming him, so that his extinction appears to be a matter of time; for no Government would face the rare opportunity which would be afforded for misrepresentations by taking steps to protect so interesting a beast from extermination. Pity it is that he must disappear, and with him one of the greatest charms of forest life, and also a form of sport that has been not only enjoyable, but beneficial, to hundreds of exiles."[51]

2
People and Tigers: Together in Life and Death

"It takes two to make one brother."—Israel Zangwill, *The Voice of Jerusalem*, 1921, 89

Baudesson recalled some Muong once catching a tiger in a pit-trap set for deer. The tiger was unhurt, but stuck fast. What happened next was witnessed by one of Baudesson's colleagues; after he agreed to leave his rifle behind. For far from setting out to kill the tiger, at great risk to themselves the villagers called on all their ingenuity to free it unharmed. First they built a cage without a floor and lowered it over the tiger. They somehow then managed to pass ropes through the bars and under the tiger's torso. Finally they hauled the tiger out and let it go; while offering it their sincerest apologies for detaining it for so long.[1]

Why? Because throughout Asia it has long been believed that tigers have spirits or souls, just like people—human souls, even—and the Muong were terrified that if the tiger died its spirit would haunt them ceaselessly.

People and tigers, it is said, were once siblings. Being such a powerful and awe-inspiring creature, the tiger is clearly a most eminent relative, one commanding the utmost respect; but it is clearly also an estranged one.

One story from Java, cited by American anthropologist Robert Wessing, has it that tigers and people once lived under the same roof in civilized vegetarian harmony; until one day the man preparing lunch cut himself and accidentally garnished it with a few drops of blood. The tigers liked the taste so much that they then went their own way as flesh-eating wild animals.[2]

In Aceh in northern Sumatra in the early 1980s, a Gayo man told

Wessing that, long ago, a man making love to his wife in a hut in a rice paddy one afternoon heard footsteps, hastily withdrew, and ejaculated. Some of his semen landed on the floor, where it became the first cat, and some shot through the doorway onto the ground, where it became the first tiger. So the tiger was fated to live outside its father's home. In another version of the same tale, told to the anthropologist in Java, the man was Sayyidina Ali, the prophet Muhammad's son-in-law and an Islamic symbol of strength — in Middle Eastern tradition he is associated with the lion — and some of his semen also fell in a river, where it became the first crocodile. (As Wessing noted, the tiger and crocodile are associated with Shiva, so probably this was originally a Hindu tale, before Islam supplanted Hinduism as the main Malay religion in the sixteenth century.) In a third version, told by the Kerinci people of the mountainous western part of Jambi province in central Sumatra to Dutch anthropologist Jet Bakels in the early 1990s, the tiger initially lived amicably in its father's home, but as it grew up it began killing his livestock, so he banished it to the jungle.[3]

In another story from Aceh, recorded by Dutch medical officer Julius Jacobs in 1894, Sayyidina Ali fell asleep on a rock with a vulva-shaped groove. While he slept he had a wet dream and ejaculated into the groove, from which a large boy later emerged. Fearing he was a young giant, people thrashed him with sticks until he went on all fours and was striped with scars: the first tiger.[4]

Around the same time, Walter Skeat, a civil servant in the state of Selangor on the west coast of the Malay Peninsula, heard that an old man once found a strange boy, with white skin, green eyes, and long nails, in the forest, took pity on him, and adopted him as Muhammad the Orphan. At school the lad was disruptive, so the master beat him with his cane. At the first blow he leapt to the doorway, at the second to the ground below, and at the third into the grass. At the fourth he began to growl, and at the fifth, now scarred with stripes, he grew a tail and went on all fours. "This is of a truth God's tiger!" cried the master, banishing him to the forest fringes.[5]

John Hutton, a district officer in the Naga Hills in Assam, reported in 1921 that the Sema Naga tribespeople there say that long ago a man, a tiger, and a spirit were brothers, born of one woman. When the man or spirit looked after their mother they washed her, cooked her rice, and gave her rice beer, so she kept well. But when the tiger looked after her he scratched her and licked up her blood, making her ever weaker, until one day she told the man and spirit, "I am going to die today. Let the tiger go to the fields. When I am dead bury me and cook and eat your meal over my grave." They did as she told, then the tiger came looking for her, crying out, "Where is my mother?" He scraped about but failed to find her, and at last fled into the jungle.[6]

Angami Nagas, wrote Hutton, say that the mother, fed up with the man and tiger's squabbling, told them to race each other to a marker, the loser to be cast out into the jungle. Knowing the tiger could outrun him, the man got the spirit to set the marker quivering with a well-aimed arrow just before the tiger reached it. Thinking he had lost, the tiger went away growling into the jungle. The man then sent the cat with a message for him: "After all, you are my brother; when you kill a deer, please put a leg on the wall for me." But the message the cat passed on was this: "When you kill a deer put it on the wall for the man." The tiger hated man ever after; but also feared him, having seen him lift rocks as big as a basket.[7] (Tigers and cats are old enemies in Malay folklore. Locke heard Malays talk of the cat as Che Gu — "Mister Schoolmaster"— because it came into the world before the tiger and was told to teach the newcomer all it knew. This delighted the tiger, and for a while the two were best friends. But the cat craftily did not teach the tiger how to climb trees, and when the tiger realized this it flew into a rage and vowed to kill the cat, which fled to the sanctuary of the human home, and has covered its tracks by burying its droppings ever since.[8])

The upshot of the various shared-ancestry tales is that the tiger, being unruly, had to leave the family home. Man ruled his world, while the tiger became Lord of the Forest — or, more accurately, of its fringes, on the boundary between civilization and the truly wild — and the two agreed to leave each other alone; with, as Wessing points out, priests or shamans — the link between people and ancestors that ensures the continued prosperity of society, and represented in the Naga stories by the spirit — mediating between the two.[9]

If a tiger killed one of their cows, reported Captain John Butler in 1854, the Angami Nagas would place an egg on the spot and call out, "O spirit! Do not, we entreat you, kill our cattle from this day forth. This is not your residence, your abode is in the woods, depart hence from this day."[10]

In most places — Java is one exception, according to Boomgaard — the occasional loss of livestock has traditionally been seen as reasonable payment for tigers keeping crop-raiding deer and wild boar in check, or as simply the will of God (or, as in the Malay Peninsula, reported Locke, of Allah), or the gods. Boomgaard noted that some people in Sumatra even used to put out wild boar carcasses to attract tigers and so keep crop-raider numbers down, while in Java some villagers would put out meat every day for their own *macan bumi*, or "village tiger," so that it left their livestock alone and kept other tigers at bay.[11]

Some people claim a totemic ancestral link with the tiger. India has numerous tiger septs, or clans, especially among tribespeople, and low Hindu castes of tribal origin, like the Pankas, a Dravidian caste of weavers and chowkidars. According to Swedish missionary and ethnologist Per Juliusson, members of the Wagmarenicka Gond tiger clan—*wag* being a variant of *bagh*—claim descent from a prince who was raised by a kind tiger but who one day, to his eternal regret, lost his temper and beat it with a bamboo; leaving it scarred with stripes.[12]

Presumably this was Singbaba, the prince in *The Song of Sandsumjee*, a Gond fable that inspired Rudyard Kipling's *The Jungle Book*. As recorded in verse form by ethnologist Robert Latham in 1859, then in prose form in 1877 by Robert Sterndale, the ruler Sandsumjee had six wives, but no heir, so he married a seventh, who bore him Singbaba, or "lionboy," while he was away on a long journey. Furious, the other wives put a puppy in Singbaba's place while his mother slept, then threw the infant in front of some buffaloes. But instead of trampling him, the buffaloes suckled him, and when the six wives returned they found him playing happily; so they threw him in front of some cows. The same thing happened; so they threw him down a well. When that did not work either, they threw him in front of a tigress, and returned home confident they were rid of him once and for all. But the tigress took pity on the crying baby, saying, "It is my child," and suckled him in her den with her cubs; much as a she-wolf suckled Romulus and Remus in Roman folklore.

One day Singbaba said to his new mother, "Clothe me, for I am naked," so the tigress sat by the road until some clothmakers came by, at which she roared and charged, making them drop their bundles and flee in terror. Singbaba clothed himself from the bundles, then knelt and kissed her feet; the greatest sign of devotion an Indian can show. Then one day he said to her, "Get me a bow," so again she sat by the road, until a sepoy came along. Again she roared and charged, making the man drop his bow, which Singbaba used to shoot birds for his tiger brothers to eat.

When Sandsumjee finally came home, Singbaba took his new family there. Among the gathering was a Brahman. Singbaba told him to rise, but he refused, so one of the bigger tiger brothers ate him. Everyone then asked of Singbaba, "Who are you?" and he replied, "Ask the buffaloes." They did, and the buffaloes told their story. "Now ask the cows," said Singbaba. The cows told how the six wives threw him down the well. "Now ask my mother," said Singbaba, indicating the tigress. When the tigress had told her story, everyone fell down and kissed her feet, worshipping her as a goddess; then gave her the six wicked wives to eat. And so Singbaba was restored as his father's rightful heir.[13]

In another version of the Singbaba legend, told to folklorist Durga Bhagvat by a Baiga, a Bhil woman gave birth to a baby boy while gathering firewood, and put him in a basket. By and by a tiger came along and carried him off. When the boy — Singbaba — became of marriageable age his tiger father fetched him a princess. Every tiger in the jungle came to the wedding. The next day Singbaba went out hunting, but fell asleep under a tree, allowing a jealous barber to creep up and murder him. When Singbaba failed to return home, the princess assumed the worst and returned grief-stricken to her father's palace. The dastardly barber then turned up claiming to be Singbaba, and somehow fooled the king. Meanwhile Singbaba's tiger father found Singbaba and magically restored him to life. Singbaba marched straight to the palace and reclaimed his overjoyed wife. Coming to his senses, the king had the barber put to death and declared Singbaba his rightful heir.[14]

As told by eminent German-American anthropologist and orientalist Berthold Laufer in 1912, in Chinese legend Jo Ngao (789–763 B.C.), prince of Ch'u, married a princess of Yun, who bore him a son, Tou Po-pi. Shortly after, Jo Ngao died and his widow returned to Yun with the boy. A few years later the lad had an affair with a court princess who, unknown to her husband, bore Tou Po-pi a son. Tou Po-pi's mother ordered the baby to be abandoned in a marsh, where a tigress found and suckled him, raising him as her own. One day the cuckolded prince saw the tigress and child together. When he told his wife, she confessed all, but he forgave her and fetched the child home. They named him Nou Yu-t'u — "Suckled by a Tigress"— and he grew up to be minister of Ch'u.[15]

Indian tiger-clan members would never countenance harming a tiger. Ethnologists Robert Russell and Rai Bahadur Hira Lal reported in 1916 that if Kathiota Kol tiger-clan members even heard of a tiger being killed they would smash pots, shave their heads, and distribute food in mourning, just as if a human relative had died. They believed no tiger would ever harm them, and if they saw one they bowed down before it in puja, or worship, even claiming that if they met a tiger in the jungle they would salaam it, and say, "Maharajah, let me pass," and the tiger would always do so.[16]

In West Java, reported Dr. Christiaan Snouck Hurgronje in 1906, legend has it that Sheikh Mansur, son of Ageng Tirtayasa, the great seventeenth-century Sultan of Banten, once freed a tiger from the grip of a giant clam, "in return for which he could always command the services of tigers, and neither he nor his descendants were ever molested by them." So strong is the feeling of kinship and common descent in West Java, noted Wessing, that all anyone there had to do if they met a tiger was say, "Don't be bad to us, who are the grandchildren of Sheikh Mansur, who is buried in Cikadueun," and they were free to pass.[17]

Some individuals claim a special relationship with tigers. Wessing cited a report that in the Dutch-Acehnese war at the end of the nineteenth century the crown prince of Aceh was alone and starving in the forest when a tiger came and licked him. The prince screamed and the tiger walked away, but when it returned the next day he realized it was a gift from Allah. Belief in such "follow tigers" was widespread on both sides of the Strait of Malacca. Even some colonials became convinced. In 1929, American traveler Mary Hastings Bradley recalled a German planter in Sumatra telling her, "I used to laugh at that as native talk. Now I know better, for my sons have 'follow tigers' that see that they come to no harm."[18]

Locke was more sceptical. Hearing of Malays who fed tigers by hand, he repeatedly called at one such man's house in Cherating in the state of Pahang, south of Terengganu, but he was never there: a classic example of what has been dubbed the Eldorado Syndrome.[19]

An old *penghulu*, or headman, in Terengganu once told Locke and a gathering of Malays — all of whom lapped up the story — that when he was a boy a man who lived alone on the edge of the jungle was so friendly with tigers that visitors often saw two or three lounging around outside his home, where he left food for them every day. His favorite was an old male, which would accompany him into the jungle and watch over him there. One day, a herd of seladang — a kind of wild ox — led by a great bull trashed his rice paddy, so he called up his favorite and showed it the bull's tracks. The tiger set off in hot pursuit; only to return two days later, exhausted, muddied, and bedraggled, with a gaping wound in one shoulder. The man summoned some helpers, including the headman's father, and the tiger led them through the jungle to the scene of an obviously titanic struggle, where the great bull lay dead, then sat and watched as they butchered the carcass. "And do you know, *Tuan*," said the headman, "that my father saw the tiger's master walk up to it and offer it a piece of meat, but it refused to eat and only pressed its body against the man's legs."[20]

The tiger also commands respect as an agent or even manifestation of powerful, often forest-dwelling, spirits or deities. In the bulging pantheons of tribespeople in India these typically include the Earth Mother, various tiger deities, and several mainly malevolent deities borrowed from their Hindu neighbors, notably Shiva and his consort Kali, usually in the form of the fierce, tiger-riding Mahadevi, or "Great Goddess," Durga, or Bhawani. Kali, the Earth Mother, and tiger goddesses were often identified with each other, as were tiger gods and Shiva. Austrian Catholic priest and anthropologist Father Stephen Fuchs, who lived in Mandla in the 1940s and '50s, observed that in the October Hindu festival of Dasara, when effigies of Durga are submerged in rivers, Gonds there made an effigy of their tiger god Bagheshwar, sacrificed a goat or pig before it, then submerged that too.[21]

In 1894, folklorist Sarat Chandra Mitra recalled hearing that in Gaya in Bihar tigers came at night from the Barabar Hills on the other side of the Punpun River to a temple dedicated to a tiger goddess and touched their foreheads to her effigy's feet, beseeching her to allow them human victims. To counter this, people held an annual fair at the temple, beseeching her not to listen to the tigers.[22]

Until the British put a stop to it in 1853, reported Russell and Hira Lal, Gonds sometimes offered human sacrifices at temples dedicated to Kali, for protection against tigers. They would lock a victim inside overnight and by morning he would be dead, all the blood drained from his body, supposedly sucked by Kali as a tigress. In Bastar, Gonds venerated Danteshwari, or "She with the Teeth," a goddess identified with Kali. Early in the nineteenth century, to protect himself from tigers on a journey to visit the Rajah of Nagpur, the Rajah of Bastar reportedly burnt alive twenty-five men at her temple.[23]

Forsyth noted that Gonds built Bagheshwar a hut a little way off in the forest, keeping him out of the village yet near enough to watch over them. The Kusru, Duri, Markam, Netia, and Sarsan Gond clans all claimed Bagheshwar's special protection. Mitra outlined their version of the god's origin. Kusru, Duri, Markam, Netia, and Sarsan were brothers. Kusru's wife bore a tiger cub, which both parents loved as a son. When Kusru watched his crops, his tiger son watched with him. One day a sambar destroyed the crops, and when the tiger child saw Kusru weeping, it tore the deer to pieces. Not long after, the tiger child died. Then, at Kusru's daughter's wedding, a *bhut*, or *nat*—a spirit or ghost—possessed one of the party. A Baiga priest quizzed it, and pronounced it the bhut of the tiger child, which demanded offerings. When offerings were duly made, the bhut departed.[24]

Ever since, Bagheshwar has possessed someone at a clan wedding, as witnessed by various colonials, including Elwin. On May 2, 1933, the Oxford University educated pastoral officer wrote in his diary: "Very exciting wedding in village. Bridegroom is member of Tiger Sept, that is, he has special relations with the tiger world and is under its protection. In middle of proceedings, which even normally bear a closer resemblance to the O.U.D.S. [Oxford University Dramatic Society] smoker than anything else, four [adult male] celebrants become possessed with spirit of tiger. At once they begin to roar in alarming fashion, jump in the air and bounce about on all fours. Presently they leap on a goat and, tearing open its throat with their teeth, drink the hot blood."[25]

Late nineteenth-century district officer and folklorist William Crooke reported that Tharus and other tribal herdsmen and woodcutters of the Terai—the belt of marshy forest between the Himalayan foothills and the Ganges plains—revered Bansapti Ma, "Mistress of the Jungle," who protected them from leopards and tigers; so long as they sacrificed livestock to her. Incur her wrath and she would set a leopard or tiger on you, or take the form of an avenging one herself.[26]

In the Ganges delta region of Sundarbans, where tigers still take their toll of people who fish, cut wood, and collect honey and other produce there, villagers say the man-eaters are controlled or possessed by the owner of the delta's riches, the tiger god Daskin Ray. Those killed and eaten have inadequately propitiated him or his consort, Bonobibi, or have abused the privileges granted by them. According to Mitra, people there even believed that, to stalk their victims undetected, the tigers synchronized their footsteps with the chops of a woodcutter's ax.[27]

The Muong made offerings to Duc-Thay, "the Lord Tiger" or "Noble Master." Even to cut down a tree without his blessing would incur his wrath. Surveying some land one time, Baudesson ordered his Muong coolies to fell a tree blocking his view. The "foreman" of the gang stepped forward and called out: "Spirit who hast made thy home in this tree, we worship thee and are come to claim thy mercy. The white mandarin, our relentless master, whose commands we cannot but obey, has bidden us to cut down thy habitation, a task which fills us with sadness and which we only carry out with regret. I adjure thee to depart at once from the place and seek a new dwelling-place elsewhere, and I pray thee to forget the wrong we do thee, for we are not our own masters." He then made a similar address to Duc-Thay. Baudesson often saw old Muong men and women with notched lathes around their necks, each notch representing a chicken or goat promised as a sacrifice to appease the deity. If they failed to fulfill the promise they simply extended it by promising an even bigger offering, such as a pig.[28]

In Manchuria (northeast China), wrote naturalist Richard Perry in 1964, people reckoned their tiger god, the Great Van, or Le Grand Vieillard — "The Great Old Man" — demanded regular human sacrifices, so in winter, when the ground was frozen, instead of burying convicted thieves alive they tied them to trees as offerings to him; and sure enough tigers would eat them.[29]

Some say Daskin Ray was an eminent ancestor killed by a tiger. Likewise, noted Crooke, many Gonds revered Dulha Deo, a boy deified when a tiger killed him just before his wedding, and Gansam Deo, a chief deified when a tiger killed him just after his wedding. As well as appearing before many of his old friends and persuading them to worship him in exchange for protection from tigers and other dangers, Gansam's restless spirit is said to have impregnated his widow. Meanwhile, at Petri in Berar in central India, Gonds worshipped a tiger goddess at a spot where a woman vanished into thin air on being seized by a tiger, as if rescued by some supernatural agency. Some Gonds, noted Mitra, even saw Bagheshwar as simply the concentrated essence of the spirits of all Gonds ever killed by tigers.[30]

In the Malay world the spirits or souls of illustrious ancestors — rulers, holy men, culture heroes, and shamans — are widely said somehow to manifest themselves post-mortem as tigers, in which form they generally continue to act as guardians of their people. Ever since Indonesia's first president, Sukarno, died in 1970, people say they have seen him in both tiger and human form. In 1974, heard Wessing, a tiger said to be Sukarno walked into a school in Yogyakarta in Java one morning and spent the day wandering around the classrooms. Later he was seen in his human form with his trademark swagger stick and white shirt.[31]

Best recalled an old Baiga shikari friend, Amoli, saying that only the best men somehow become tigers or elephants in the next life, but not everyone agreed. In 1928, in a story that features "purely fictitious" characters but was evidently based on real events, India born and bred Maurice Hanley, a tea-planter in Assam, told of a devilishly cunning tiger that preyed on coolies on a tea estate at Khagrijan. The coolies said it was the spirit of an old man, Karamsing, come as a tiger at his widow's bidding to avenge all who had crossed the couple. They spread this story after the tiger killed a man who had quarreled with the woman. In other words, they used the man-eater as an excuse to get rid of her. Only the intervention of the British estate manager stopped them from tearing her to pieces.[32]

Fuchs reported that the Baigas and Gonds of Mandla believed a bad man who died unrepentant could not find peace and came back as an *olthwa*, a tiger-form creature that sucked your blood without killing you. One was even seen sitting in a tree with bangles on its legs and rings in its ears: ornaments that proved its human origin.[33]

Exactly how people become tigers post-mortem depends on who you talk to. Worldwide, post-mortem metamorphosis is widely reported, yet shrouded in mystery and confusion. Sometimes the spirit or soul is said to take the ghostly or even physical form of an animal, and sometimes it is said to enter and control a real animal, a process known as transmigration, or metempsychosis. Both processes are distinct from reincarnation, which is rebirth of the soul in another body, human or animal. What all three have in common is that no one can disprove them: unlike another process sometimes reported, namely the physical transformation of the corpse into a living animal.

In the Malay world, one way of becoming a tiger post-mortem is a kind of transformation involving a cricket, a prominent soul-carrier in Malay magic and one of various small creatures Malay witches keep as bloodsucking *pelsits*, or familiars. Wessing cited a 1968 account of a man from Sunda, in West Java, who was buried immediately, according to Islamic custom, and that evening a noise was heard coming from his grave, in which a small hole appeared. Out of this came a wisp of smoke and a cricket, which grew into a *bobongkong*; a doll-like creature, half man, half tiger, that walked on tiptoes. But in another account from the same source a man's corpse was laid out with a lamp alongside it, and that night the mourners keeping vigil saw a cricket jumping around the light. As they watched, it turned into a cat, which walked around the corpse seven times before turning back into a cricket and disappearing. To their alarm the corpse then grew a tail and began turning into a live tiger. They started yelling and the metamorphosis halted (and presumably conveniently reversed itself). Likewise, reported Wessing, in Aceh it is said that moments after holy man Teungku Chik di Rakmeh died his face changed into that of a tiger, then three days after he was buried his grave lengthened as the metamorphosis extended to his body. (Presumably the grave was later conveniently found to be empty.)[34]

In 1891, John Lockwood Kipling, father of Rudyard, wrote, "You can be shown to-day [in India] forest shrines and saintly tombs where the tiger comes nightly to keep a pious guard." Likewise the graves of eminent ancestors in the Malay world are widely said to be guarded by tigers that are the transformed souls or bodies of the graves' occupants, or real tigers containing the occupants' transmigrated or reincarnated souls. Wessing noted that, despite the disapproval of the authorities, many people in Aceh claimed to have seen tigers by such graves, especially in the evening. But not everyone is convinced they are supernatural creatures. A Gayo shaman told Wessing that while he did once see a tiger lying under some trees near such a grave, it could just have been an ordinary tiger sheltering from the rain. As Boomgaard points out, such graves tend to be in wild, remote areas where, as sacred spots, they can become wildlife sanctuaries highly attractive to tigers.[35]

Some say that tigers that guard graves are the ancestors' ghostly servants. According to Hurgronje in 1903, legend has it that when the ghost of the Gayo holy man Teungku Ulu-n-Tanoh sent his tiger with a delicacy for the ghost of a colleague, Teungku di Kuto, the tiger traveled so fast that the tidbit was still warm on arrival, despite their graves being a seven-day walk apart.[36]

In Sunda, tiger-form ancestors are said to be afoot in the month of Mulud, which marks the birth of Muhammad. If you have a problem you can ask them for help at home or at their gravesides. Wessing heard that a man facing trial for hiding a gun for someone fasted by an ancestor's grave for three days. A tiger appeared and pinned him down with its claws, but he begged forgiveness and pleaded for help. The tiger vanished and the man was later acquitted. The anthropologist also heard that when a man lost some government property and asked for help at an ancestor's grave the items materialized before him.[37]

In a time of family problems, tiger-form ancestors in Sunda may even turn up at night uninvited to offer help. Likewise, when American anthropologist Kirk Endicott lived among the Batek Negrito hunter-gatherers of the Malay Peninsula in the early 1970s, in the east coast state of Kelantan, they told him their ancestors occasionally visit them — for real or in dreams — in tiger form to see what they are up to and to teach them useful things like songs. A Batek may sometimes also have a chance encounter with a tiger-form ancestor in the forest, in which case he blows incense smoke over it to temporarily change it into its human form. Many Batek carry incense on them for just such an eventuality. Dreaming can also be dangerous, however, for that is when your "shadow soul"—the African equivalent is your "bush soul"—wanders free and might be spotted by an ordinary tiger, even in the dark: the tiger might then follow it "home" and kill you in your sleep.[38]

In Central Java, some say that those born at sunrise in the first month of the Javanese year can expect lifelong protection from a tiger-form ancestor. Wessing heard of one such baby that was put outside with a bowl of flowers in water when it fell ill. The baby's guardian ancestor came in tiger form, dipped its mouth in the bowl and licked the baby all over, after which the baby got better.[39]

The Malay world abounds with stories of ancestors acting as tiger-form familiars to their descendants. All the inhabitants of a coastal area of Aceh identified as the old kingdom of Daya are said to physically transform into such creatures post-mortem; either in the grave, or when their corpses are washed with diluted lime juice according to Islamic custom. Wessing heard of one such woman who became the familiar of her only child, Mak Sum. Whenever Mak Sum had to go somewhere of an evening she rode on her tigress-mother's back, and when she in turn died, in 1979, she too, people say, became a tigress.[40]

Malay shamans—*pawangs* or *bomohs*—typically pass on their powers and knowledge a few days after they die, noted Endicott. Most often this *ilmu* is passed from father to son, the latter waiting in the forest or by the graveside until his father appears in tiger form. The father or an earlier ancestral shaman then becomes the son's *hantu belian*, or tiger-form familiar.[41]

Shamans of tribal origin in Malay countries acquire their knowledge and powers in much the same way, their tiger-form ancestors likewise acting as familiars. When Bakels stayed among the Kerinci in the early 1990s, their *dukuns*, or shamans, told her they can ride on theirs—like Mak Sum—and order them to guard their fields.[42]

Various witnesses have written about the *berhantu*, a séance in which, to cure a patient, a Malay shaman makes offerings like rice, eggs, betel and scented oils, including some for Shiva in his role as divine healer, and is possessed by his hantu belian. Endicott has studied these reports in some detail, and typically the cure takes three nights.[43]

One account missing from Endicott's study, however, is that by

2. People and Tigers

Malaya-born, Kuala Lumpur-based investigative Radio Malaya reporter Stewart Brooke-Wavell (Figure 11), writing as Stewart Wavell, in 1958. Recording a Hindu shadow-play one night, Wavell was just admiring a young Thai woman a few rows in front of him when she fainted. Her Malay husband could not lift her so Wavell gallantly stepped forward and picked her up. The husband led the way to the dimly lit house of the local pawang; an old man whose eyes, recalled Wavell, bored into his own with an embarrassing acuteness. The three men laid the now conscious but dazed girl on a mat and slipped a cushion under her head.

With her husband watching meekly, the pawang removed her sarong and underclothes, talking all the while. Wavell could not follow everything he said, but gathered that she was called Aminah and was frigid, having received some sort of shock as a child. Now she lay naked on the floor. "She was finely made," observed Wavell, "and the husband's sense of frustration was understandable."

Covering her lower half with her sarong, the pawang hurried around the room, gathering various items and muttering to himself. Then he spoke to Wavell: "Rarely do this nowadays. Muslims object to ancient Hindu customs. Aminah dance like Kali. Only hope. Get rid of evil spirit inside. Then she get better and make love. Husband very happy. You help, good?"

Remarked Wavell: "I nodded, without fully understanding what he had said. Then with appalling clarity the sequence struck me." Meanwhile the husband sat in the corner, looking understandably worried.

Waving lit joss-sticks, the old man muttered and whistled, then paced the floor, scattering rice and calling to various Hindu deities including "Kala"—"Black," or "Death," meaning Kala Bhairava—one moment pleading softly, the next shouting a torrent of abuse. Then he tore off his clothes and began leaping about, while Aminah watched, glazed-eyed. Once he almost landed on her stomach, but Wavell pulled her clear just in time, dislodging her sarong. She looked at him in puzzlement, but made no attempt to cover herself up again.

Now on all fours, the pawang padded around the floor, baring his teeth and tongue and growling. Wavell considered slipping away, but the thought of being alone in the dark with a "human-tiger" on his trail decided him against it.

Purring—though tigers cannot actually purr—the pawang leant over and stared into Aminah's eyes. Then: "Slowly, almost imperceptibly, he lowered his head, her eyes watching him, and very delicately he licked her stomach, at first cross-wise, then downwards, down towards the centre of her fear. She quivered under his touch, her legs spasmodically folded and opened.... The girl's breathing quickened. Her breasts were sharp and firm points of expectancy. She began to move with a timeless rhythm...."

At last he stood up, drawing her with him. She began to dance provocatively, and Wavell barely noticed the pawang slip away. Her eyes met his, and she smiled invitingly. But Wavell waved her husband over instead and went outside, where the pawang thanked him for his help. "Tonight," he said, "you have been the instrument of Kala."[44]

Malay tiger-form ancestors demand due respect. In Banda Aceh, Wessing lived near the grave of holy man Teungku di Krueng, who is said to guard it in tiger form. When some soldiers once held a party nearby and things got out of hand, the tiger chased them away. In Sunda, if you have any kind of ceremony you must invite your tiger-form ancestors and leave them offerings like tobacco outside, lest they be insulted and disrupt proceedings. And you must follow *adat*, or custom: the sacred rules laid down by the ancestors. A Sundanese man told Wessing how a tiger-form ancestor possessed him when he held a joint circumcision ceremony for two of his sons: "I went off by myself toward the back of the house because the crying upset me suddenly. When I returned I was *lupa* [crazy] and scattered the feast. When I came to I had raw meat in my mouth."[45]

Sometimes Malay tiger-form ancestors punish wrongdoers severely. Anyone swearing an oath on the grave of Teungku Chik Cicem in the Pidie district of Aceh, noted Wessing, must embrace one of two stones said to have followed the holy man there from the Besar district, and if they are lying a tiger will eat them. Boomgaard cited the case of a woman in Banten who lost her three children to a tiger one day in 1839; people said that it was because a European had desecrated Mount Dangka, the ancestral spirits' home. The Batek told Endicott that if someone failed to make offerings as thanks for the first wild fruits of the year, then either their thunder god, Gobar, would send a storm to kill him — by lightning strike, or by having a tree crash down on him — or a tiger-form ancestor would hunt him down. Or perhaps one, or both, would have him "accidentally" fall out of a tree. Skeat heard that the normally harmless ghost tiger guardian of a Muslim shrine in the Kuala Langat district of Selangor killed a Chinese coolie from the local pepper plantations for leaving it an offering of pork. Such tigers were said to have one foot smaller than the others. Locke heard that they first killed a cat when their wards did any great wrong. If no heed was taken of this they killed livestock: and if that failed they killed the offenders.[46]

2. People and Tigers

Tigers that persistently take valuable cattle have long widely been considered evidence that a tiger-form ancestor or other deity needs propitiating. In 1887, on the border between the Chhindwara and Seoni districts of central India, Hicks rid the Gond villagers of Konapindrai of an old cattle-lifting tiger that lived in the nearby Dongardeen caves in a bend of the Pench River. The forest officer led a party of Gond beaters there, and they flushed out the tiger, but try as he might he could not get a clear shot. That evening he gave them the means to propitiate various deities, plus a few rupees for the local Brahman priest, to secure the man's prayers and blessings. The men returned with him to the caves the following morning in high spirits, confident of success. Wrote Hicks: "It was amusing to watch these Gonds as they jumped from rock to rock, laughing and chattering like children, peering and throwing down fire-works at the tiger below them, saying: 'here is another one for you! come on out and be done with it, for we have done *puja* to the *Barra-deo* [Great god] and he has given you to us!' In this manner, accompanied with much personal abuse regarding himself and his ancestors, the tiger was finally driven into a corner from whence there was no escape."[47]

Hicks also recalled a cattle-killing tiger in Chhindwara that was so long-lived — forty years, people said — that many people had become rather fond of it, holding it in supernatural awe because it had survived being shot by the former deputy commissioner. They were reluctant to help anyone hunt it, and sportsmen and shikaris alike left it alone. But it caused grazing license revenues to fall, and finally it accidentally killed someone, so in 1887 the district magistrate asked Hicks to shoot it. Hicks soon discovered that a group of Brahman priests had deified it, making people afraid to help him. (Brahman priests were renowned for their "arrogance, greed, hypocrisy and dissimulation," noted Russell and Hira Lal.) So Hicks bribed a rival group of Brahman priests to intercede on his behalf, and soon had plenty of helpers. After shooting the tiger — the biggest he ever saw — he gave the villagers all they needed "to make merry that night," he wrote, "and with which to feed and propitiate their jungle deities with thanksgivings to their hearts' content."[48]

Sanderson similarly recalled an outsized cattle-lifter in Mysore known fondly to locals as Donnay, "the cudgel," because it never hurt anyone over the years. So long-lived was the tiger, wrote Sanderson, that it had come to be regarded as enjoying the special protection of Koombappa, "the great jungle-spirit," people saying that when Koombappa "went the rounds," he rode on Donnay's back. The villagers had even made an effigy of the tiger, "respectably got up in wood and paint, and looking truly formidable, with a seat on the back, and on wheels, which they dragged round the temple and down to the river in solemn procession on feast-days." When they told Sanderson that Donnay could never be shot, he determined to prove them wrong and kill "this notable rival." This turned out to be easier said than done. "Never had a tiger so many lives, never did one retain his skin more cleverly," wrote the sportsman. But finally he did shoot it dead, at which one of his trackers said reproachfully, "He never hurt any of *us*."[49]

According to Russell and Hira Lal, some Gonds saw cattle-lifters as punishment for their wives' infidelity, while in 1962 American ethnologist Gordon Young reported the same thing of the Karens and other hill tribes of northern Thailand. But more often, it seems, the price of sexual transgression is death, by tiger or lightning strike. Malaysian anthropologist Iskander Carey reported in 1976 that the Kensiu Negritos of the west coast state of Kedah in the Malay Peninsula traditionally say their thunder god, Karei, administers such punishment not only to anyone committing incest but to any married couple who — as in the Gayo tiger-origin story — have sex by day. According to Hutton, some Nagas said that their deities punished "irregular intercourse" in the same way. Wessing noted that in Java and Sumatra it is Muhammad who sends tigers to punish sexual transgressors: in such cases, say some, the tigers cannot be shot, nor the prophet placated. William Baze, a farmer in upland French Indo-China before World War II, recalled a tiger that preyed on the young women of a Muong village. The village priest declared the tiger "inhabited" by a dead husband. Wrote Baze: "That was enough for the Moi. A wave of marital fidelity immediately swept through the village..." Less drastically, Bakels learned that if a tiger merely shows itself near a Kerinci village it means that someone has committed adultery or some other transgression. Pressure is then put on the suspect — there always is one — to confess and make amends.[50]

Clearly, the fear or threat of being killed by a tiger is a useful law enforcer and helps maintain social stability. Austrian ethnologist Hugo Bernatzik, who studied the Mlabri hunter-gatherers of northern Thailand in the 1930s, reported that they traditionally told children off by saying, "Don't do that, or the tiger will come." According to Endicott the Batek also use the tiger as a "bogeyman" to frighten naughty children.[51]

Some of the ways you can offend tiger-form ancestors and other deities — or even just ordinary tigers — seem trivial. Edgar Thurston reported that the Savara hill tribe of the Ganjam district of Orissa — now Odisha — in India said a tiger would kill anyone using any number higher than twelve: because long ago some Savaras were measuring grain and when they got to twelve a tiger ate them. Among the Kerinci, learned Bakels, all behavior linked with tigers is taboo, including eating from the cooking pot, going on all fours, taking a pestle outside — a pestle being like a tiger's tail — and even sitting with one knee held high. In Burma, Christopher's camp-followers were once about to curry a bamboo rat for their breakfast when the sportsman insisted it was not enough for a proper meal and shot a few parrots to add to the pot. They could hardly refuse, but when a tiger then appeared they feared it had come to eat them for cooking beasts of the field and air in the same pot. Fish also had to be cooked separately, noted Christopher. Endicott found that the Batek similarly forbid cooking certain foods together because it makes you give off an odor of raw meat. Drinking undiluted blood, and crushing head lice after eating frogs or squirrels, are taboo for the same reason. Break any of these taboos, say the Batek, and at best you will cut yourself badly, and at worst a tiger will attack you.[52]

Being in any way disrespectful in the tiger's forest kingdom has long been widely taboo. It was said, wrote Crooke, that Bansapti Ma would make ill or send mad anyone who dared sing or whistle in her domain, while a Kerinci told Bakels he once bad-mouthed tigers to see what would happen, and the next morning found scratch marks all round his hut. The Batek told Endicott that calling the tiger "man-killer" is particularly offensive; that tigers will kill anyone who does so.[53]

Even referring to the tiger by name is dangerous, just as, noted Mitra, it is risky to name such dangers as snakes or thieves, in case it brings them to you. Indeed, this is a worldwide superstition. Locke heard that if you spoke the word *rimau*, or tiger, then somehow, somewhere a sleeping tiger would hear you, and hunt you down when it woke. Yet many Malays freely used the word in his hearing. When he asked a pair why, they "...looked a little sheepish," he wrote, "but both gave the same answer. This was that no harm would come to anyone who used the word in my presence, as all the tigers knew of me and would do everything within their power to keep away from me and from my friends."[54]

Baker told how a friend of his once sat up on a machan for a man-eating tiger with an experienced old Muslim shikari, who suddenly whispered to him, "Look, sir." The sportsman peered hard, but could see nothing. "Look, sir," the shikari whispered again, pointing with trembling hand. "Look at what?" asked the sportsman, now somewhat annoyed, "cannot you speak?" Now shaking with terror, unable to bring himself to say the word "tiger," the shikari kept pointing and whispered once more, "Look, sir." Finally the sportsman made out the outline of the man-eater.[55]

Traditionally, people either referred to the tiger euphemistically, or gave it a completely different name, something so offensive, perhaps, that no self-respecting tiger could possibly believe they were talking about *it*. In China, noted Taiwanese scholar Man-ping Chu, people called it "Big Insect," while in India, as recorded by Sanderson and many others, people often referred to it as a dog, or a jackal, "partly in assumed contempt," noted Sanderson, and "partly from superstitious fear."[56]

When confronted by a tiger, the safest recourse is to call it something respectful. A touching example of this was reported from India in 1873 by Colonel James Tod. A tiger seized a boy near his camp one night, but the lad cried out, "Oh Uncle, let go—let go—I am your child, Uncle, let me go!," and the tiger released him. As Boomgaard points out, however, addressing a tiger as an eminent relative or ancestor is not necessarily always meant literally, for kinship titles are used widely as honorifics in Asia. In the Malay world, for instance, the word *datok* is used for grandfather as well as chief.[57]

In Java, reported Wessing, a Muslim villager entering the forest sought the tiger's blessing by appealing to their common ancestry, saying, "Embah [a term of respect], please do not hurt me. I am Adam's descendant—please step aside and do not bother me."[58]

Hubert Banner, who spent twelve years in Java, wrote in 1927 that when people there passed through jungle likely to harbor a tiger or two they laid on the flattery with a trowel, singing such fawning songs as:

> Tuan Nan Gedang, Mighty Lord,
> Hard is the wood, hard is the iron which cuts the wood;
> But harder yet than wood and iron
> Is the Mighty Lord of the strong claw,
> That tears the wood and snaps the iron of the spear.
> Hailah! Hailah!
> Who is stronger than the Tuan Nan Gedang?
> The buffalo is big, the buffalo is mighty.
> The foe dreads his horns, his horns hard and sharp,
> And his pointed hooves that trample his fallen enemy.
> But stronger than the buffalo art thou.
> Hailah! Hailah!
> Strong is man; he turns the wood, the iron he forges;
> He bends the buffalo to the yoke and compels his labour.
> But stronger yet than man, and more wily, is the Mighty Lord.

> The Mighty Lord is man's friend, and wishes him no ill.
> Hailah! Hailah!
> The mighty and the good, 'tis Tuan Nan Gedang!

Then, as soon as they were clear they spat and shouted back insults.[59]

The only people in Java who could pass safely through a tiger's territory without recourse to flattery were government officials, people for whom the tiger had the deepest respect. Boomgaard cited a nineteenth-century report of one squatting before a tiger and saying "Letter of the High Government" over and over until it let him by. Likewise in Sumatra, Sir Stamford Raffles' widow, Lady Sophia Raffles, recalled that in 1818, when her husband was governor-general of Bencoolen (now Bengkulu) province, in the southwest, some of his coolies once found their way blocked by a tiger. Supplicating themselves, they explained to it that their master would be angry if they were late, and the tiger moved off. No doubt they startled it, reckoned Lady Raffles, but they were "perfectly satisfied that it was in consequence of their petition that they passed in safety." At Panti in West Sumatra, Bradley heard that a government mailman, who wore a brass plate on his belt as a badge of office, once turned a corner and found himself face to face with a tiger. Though terrified he had the presence of mind to say boldly, "Do not attack me. I am a government servant. Here is my brass plate." The tiger recognized the badge and immediately leapt aside.[60]

Instrument of divine punishment or not, a "guilty" tiger can be called to account. In China, Qing dynasty writer Pu Pongling (1640–1715) — as cited by American author Bernhardt J. Hurwood in 1968 — told how a tiger ate the only son of a poor old widow. Distraught, she begged the local magistrate to issue a warrant for its arrest. Laughing, he told her to stop wasting his time, but she refused to budge. Finally in exasperation he wrote out the warrant and ordered an officer called Li Neng to carry it out.

A month later, despite several beatings from his boss, Li Neng still had not found the tiger. In desperation he fell to his knees, sobbing, in a temple in the forest and begged the gods to help him. Hearing a slight sound he turned around and there, in the doorway, looking straight at him, sat a tiger. Thus cornered, he resigned himself to his fate. But the tiger did not move, so with nothing to lose he walked straight up to it and said, "Oh Mighty One, if you are the man-eater, then you must allow me to place you under arrest," at which it lowered its head. As if in a dream, Li Neng found himself slipping a rope around its neck and walking it to the magistrate's office as if it were a dog on a lead.

Keeping his composure, the magistrate demanded of the tiger, "Are you the one who killed and ate the old woman's son?" The tiger nodded. "Well then, normally you would have to pay for your crime with your own life. But the man was this woman's sole support, so I will pardon you if you agree to take his place. What say you?" The tiger nodded again, and the magistrate ordered its immediate release.

The old woman was outraged, until the next morning when the tiger left a fresh deer on her doorstep and she sold it for a tidy sum. Each day from then on it brought her something, sometimes even money, and in time she saved enough to buy a fine new house. She came to love the tiger like a son and would sit stroking its head for hours on end as it sprawled at her feet on her veranda.

When finally she died, the tiger joined the mourners at her graveside, raised its head to the sky and let out a long, agonized roar; before slipping away into the forest, never to be seen again.[61]

Wessing cited a 1981 report of a Sumatran pawang who, to bring a tiger to justice, made a "cage" out of palm fronds, set it down, and casually called the animal forth, before leaving to attend to more important matters. Burdened by its guilt, sooner or later the tiger obeyed the summons, sometimes coming on its own but sometimes with an escort of fellow tigers, who wanted nothing more to do with the criminal but came to bid it goodbye. Once inside, it made no attempt to escape, though a child could break out of such a flimsy structure, and on the shaman's return it lowered its head in shame and spat in agreement when asked if it was guilty. A Gayo man told the anthropologist that such tigers are shunned by and banished from "tiger society," a gathering ruled over by a king tiger who wears a necklace of palm leaves. Thus they are condemned to prowl crazily around villages.[62]

When the Kerinci trapped a guilty tiger, learned Bakels, they treated it with the utmost respect and ceremonially buried it. But first they mounted it in a lifelike pose on a bamboo frame and danced before it, chanting:

> Oh grandfather
> Grandfather crippled tiger
> Who is the guardian
> Of the Kingly Mountain
> Oh grandfather crippled one
> Your grandchild did not behave well
> Your grandchild already got lost
> And he has to pay his debts

> Your laws have been broken
> Oh grandfather the crippled one
> The kris* of old time should not shudder
> The mirror of former days should not be blurred
> Your grandchild already has blood on his paws
> The fence has been opened, the border trespassed
> Words of the old days we are looking for
> Words, that have to be remembered
> The debt has to be redeemed
> The smoldering fire has already been quenched.[63]

Likewise, Wessing noted that when the Batak of North Sumatra killed a guilty tiger they treated it like an honored guest, laying out refreshments and addressing it as Grandfather. And if members of the tiger clan were around they would weep and put *sirih*—a quid made from betel leaf, areca nut, gambier, and limes or lime juice—in its mouth.[64]

When a peninsular Malay pawang killed a tiger, wrote Skeat, he explained to it that it was not he, but Muhammad, who was really responsible. It was then both honored and mocked at the sort of reception normally reserved for a visiting dignitary. When a rajah once bagged a tiger near Jugra in Selangor, it was paraded before the sultan propped up on all fours with a rope through its mouth, which was wedged open with a stick, and its head was jerked up and down as if it were still alive. To the beat of gongs and drums, two men pretended to fight it with their swords, then taunted it by dancing before it unarmed.[65]

Skeat reckoned the mouth was wedged open to enhance the impression the tiger was still alive, but another explanation is that the stick stopped it saying who had killed it, for unless it was a guilty one—and maybe not even then—tiger society was unlikely to let its death go unavenged. Among the Muong, reported American traveler Henry C. Flower in 1920, a public apology and perhaps a memorial stone were considered all that was necessary to prevent tigers avenging the death of one or two of their number, but "wholesale slaughter" would result in "a campaign of vengeance" by other tigers. In Java, noted Wessing, responsibility for killing a tiger was assumed by the whole community, and the actual killer always had a ruler's blessing, to protect him from being cursed by the tiger's soul.[66]

Elsewhere people took steps to deceive tiger society when they killed a tiger. Hutton noted that an Angami Naga who killed one was hailed as

---
*A wavy-bladed dagger.

a great warrior, but the village priest still declared a vacation for the death of an elder brother, and the killer would never admit his deed, otherwise all the other tigers would hear him and hunt him down. Instead he would say the gods had done it. To prevent the tiger telling the tiger god Tekhurho the truth, its head was submerged in a stream: with its mouth wedged open with a stick. Sema Nagas sometimes put a stone in a slain tiger's mouth, to stop its ghost waylaying them in the next world. They also reckoned the killer's soul was in danger from the tiger's soul if the man slept too soundly, so he was advised to sleep on an uncomfortable bed. And sometimes, among both Sema and Angami Nagas, when a man who had killed a tiger died, a dog was killed so that *its* ghost would protect the man's ghost from the tiger's ghost on the road to the afterlife.[67]

Meanwhile, when American anthropologist Frederic K. Lehman lived among the Chin people of the highlands of western Burma in the late 1950s, he learned that members of a successful tiger-hunting party would afterwards dress and act like women, to make the tiger's spirit think they could not possibly be the killers.[68]

Finally, among the great warrior and hunting Kodava caste of the Coorg district (now Kodagu) in the Western Ghats in southern India, in what is now Karnataka, a man or woman who killed a tiger or leopard was ceremonially honored at a *Narimangala*, or "tiger wedding"—*nari* actually meaning "jackal"—at which, according to Crooke, the hero or heroine of the hour was "married" to the animal's soul (Figure 3). This was quite possibly to prevent it avenging its killer.[69]

Representations of tigers are powerful charms, whether placed in doorways, painted on walls, hung round the neck or tattooed on the body, and however crude they are; Gonds camping out slept between two mud "tigers" to ensure their safety, observed Elwin.[70]

Imagine, then, how powerful parts of the actual animals must be. All parts of the tiger have long been valued not just as medicinal potions—and not just in Chinese medicine, either—but also for protection from the evil eye, animals, and misfortune generally, and as a means of acquiring the tiger's strength, courage, and supernatural powers (Figure 4).

Bernatzik recalled Jaray tribespeople in Laos jabbing their spears into a dead tiger to strengthen them, while Hindu Indians have long believed that a tiger's tail possesses the animal's strength, and if they could acquire one, observed Sir James MacNabb Campbell in 1885, would tie it around a child's neck. In China a rag soaked in tiger blood was believed to ward off smallpox and measles demons when worn around a child's neck, and

2. People and Tigers

Figure 3. One of the last "marriages" of a Kodava warrior to the soul of the tiger he has just killed was on March 9, 1873, an event so unusual even then that it made *The Illustrated London News* (on December 6, 1873), complete with this drawing by Captain Belford Cummins of the Staff Corps at Mercara (now Madikeri) hill station. On the left are British officers specially invited to the wedding. Seated under the canopy in full warrior costume, the groom, Colavanda Carriapah, receives the tributes of friends and relatives, who then carried him in triumph around the dead tiger (suspended from a bamboo frame).

when a tiger was shot there, reported Methodist missionary Harry Caldwell in the 1920s, villagers would swarm around it, soaking up every last drop in rags, even scrabbling for blades of bloodstained grass.[71]

Dutch sinologist Jan de Groot noted in 1901 that in China it was believed you could cure a fever by sitting and sleeping on a tiger skin, while Crooke and Edgar Thurston both noted that in Orissa the Juang, Ho, Santhal, and Khond, or Kandh, tribespeople swore oaths on a tiger or leopard skin, saying a tiger or leopard would kill anyone who broke an oath so sworn. Sema Nagas bit on a tiger's tooth when swearing oaths, and Hutton observed that many did this quite happily because tigers had become so scarce that the chances of one coming after them if they reneged were slim.[72]

Figure 4. In 1920 American traveler Henry C. Flower recalled that when a French forester in his party in French Indo-China, a man called Millet, shot this man-eating tiger dead, their Muong coolies begged them for its bones to make medicine, while one Muong woman even laid her baby boy on its freshly skinned body "so that he might absorb its sinewy vitality" and, as she put it, "grow to be like his father" (*Asia*, October 1920).

If they could get one, remarked Banner, Javanese men hung a tiger's penis over their beds to boost their sexual prowess. And while Mitra among many others noted that tiger fat was valued as a remedy for rheumatism and gout in India — "Many a useful friend among natives may be made by a present of a little tiger's fat," wrote Hicks — Burton reported that it was also valued there as an aphrodisiac. (When he boiled down Donnay's fat, Sanderson noted, it totaled four imperial gallons.)[73]

Burton also reported that the milk of a tigress was prized as an eye ointment, if you could get it. As Crooke pointed out, fetching it is one of the stock impossible tasks set heroes in Indian legends, (the others being fetching an eagle's feathers, water from the well of death, and the mystical cow guarded by demons), while Mitra wrote that in Bengal you praised an enterprising man by saying he would fetch such milk if asked. Locke heard that pills made from tigers' eyes cured convulsions. He also recalled blind Malays begging him for the eyes of the next tiger he shot: "'If you will only give me the eyes, *Tuan*,' they would say, 'I shall have medicine made from them and shall be able to see again.'"[74]

Many a sportsman on shooting a tiger found the claws and whiskers gone by the time he reached it. In Burma the whiskers were made into rings to ward off evil spirits, reported Christopher. In India, wrote Mitra, plucking or burning off the whiskers was said to avert ill fortune, prevent whoever killed it becoming a tiger post-mortem, or even prevent anyone assuming the animal's form, while owning them might win you the heart of any lady you set your sights on. Some Gonds and Korkus singed off the whiskers to prevent the animal's spirit haunting them, recorded Russell and Hira Lal, while others chopped them up to make poison. In Sumatra, according to one report cited by Wessing, it was said that, to avoid poisoning itself with its own whiskers, a tiger always faces downstream when drinking from a stream. One Gayo man assured the anthropologist that the whiskers on the left are the deadliest. Baudesson recalled a Muong member of his mission with a virulent fever being convinced he was the victim of tiger whisker poisoning and repeatedly coughing up the medicine the Frenchman tried to force down him. "Within four days he was dead," wrote Baudesson, "a victim to his own superstitious ignorance. The most unfortunate effect of the tragedy, however, was that it only served to confirm his companions in their belief."[75]

Other prized parts were the liver, eaten for courage, and the clavicles, or collar-bones. In Burma, reported Christopher, the latter were called "the Tiger's Strength," tigers being said to have to lick them before leaping. In India, noted Hicks, Burton, and Powell, among others, these "Lucky Bones"—also known as "Santokh," or "happiness," wrote Crooke—were mounted as charms. Baudesson observed that competition among the Muong for these talismans, which sometimes had a market value greater than that of a buffalo, was fierce and often bloody. (Sportsmen, too, sometimes kept the clavicles, as souvenirs. Those of a tigress Durand once shot adorned his desk for years.)[76]

Brown recorded that a charm made from a tiger's clavicle was said to ensure female fertility in India, while in 1832 Charles Coleman wrote of the Garo hill-people of northeast Bengal (now Meghalaya and Bangladesh): "The tiger's nose, strung round a woman's neck, is considered as a great preservative in child-birth; they aver it keeps off giddiness, and other disorders consequent on this event."[77]

Tiger parts could work their magic on your crops and animals too. In India, according to Crooke, people burnt tiger flesh in fields and cattle stalls to keep crop and cattle diseases at bay, and if your cattle did become diseased, noted Sanderson, tiger fat was reckoned an effective remedy. Forsyth reported that mahouts commonly fed their elephants pieces of tiger liver to make them more courageous (not to mention the eyes of an owl, "torn fresh from the living bird," to make them see better in the dark), while sportsman Major-General James Elliott cited the case of an Indian elephant that "disgraced itself" by bolting when it was only "slightly" mauled by the tiger being shot dead at its feet: that evening it was rubbed all over with the tiger's fat, and its nerve never failed it again.[78]

Representations of tigers and tiger body-part amulets are not the only safeguards against attack by tigers and other animals. In southern India, reported Edgar Thurston, the Paliyan tribespeople of the Palni Hills planted four jackals' tails in a square for protection, especially from leopards. Even if a leopard entered the square, it could not harm them, as its jaws were locked. A Javanese kris that Wessing once bought was examined by two separate parties and pronounced an effective tiger deterrent. In 1912 Mitra wrote of a certain tune — played on a *kerotong*, or two-stringed bamboo harp, and known only to a few old people — that lulled tigers to sleep in the Malay Peninsula. Also in the peninsula, just an inch of *los* or '*tas* — the wood of the master's cane in the Muhammad the Orphan tale — was said to be enough to make any tiger cower. A Malay headman once mistook Skeat's gorse walking stick for such wood, begged it off him, and cut it up to share among his followers.[79]

Then there are spoken charms. Locke noted that before entering the forest many Malays recited passages from the Koran to prevent tigers attacking them, while following an expedition to Terengganu and Kelantan in 1899–1900 — the Cambridge Expedition, as it became known — Skeat and fellow civil servant Charles Blagden recorded several spoken charms used by peninsular Malays and tribespeople to ward off tigers; though whether these charms were the preserve of shamans, or could be used by anyone, is unclear. This one, from a Blandas tribe in Kuala Langat, is for crippling a tiger:

> Trong wet! Trong wau!
> Stick fast i' the tree-stumps, where thou prowlest;
> The weighting charm is said already.
> Refuse thou then men's heads, O Tiger.
> And be your hind-feet slow, earth-loaded,
> And slow, stone-loaded, be your fore-feet.
> A sevenfold rampart now surrounds me,
> The weighting-charm I've just repeated.

And this is a Malay charm for shutting a tiger's jaws:

> Madam Ugly is the name of your mother,
> Sir Stripes the name of your body.

> I fold up your tongue and muzzle your mouth;
> Let the twig break—break with the weight of this well-fed wild goose.
> Be (your mouth) shut fast and locked.
> If a bachelor loses his vocation, it does not matter.

This was followed by a few words of Arabic. On reaching home you had to remember to unlock the tiger's jaws by repeating the Arabic words then commanding, "*Buka!*"—"Open!"[80]

Finally, Baudesson recalled that as the Muong thought all writing and paper magic, Muong couriers carried unused envelopes to protect them from tigers and evil spirits. One can only hope these brought them more luck than whole bags of mail brought runners in India.[81]

In 1784, British East India Company official William Marsden wrote that people in the Rejang area of Bencoolen talked of a place in the neighboring Pasemah area "where the tigers have a court, and maintain a regular form of government, in towns, the houses of which are thatched with women's hair." Skeat and Locke, among others, reported remarkably similar stories from the Malay Peninsula, where folklore has it that tigers actually live in human form in their own settlements, and that not only are the roofs of their houses thatched with human hair, but the walls are covered in human skin and the posts are made from human bones. Each morning the inhabitants swim across or pass under a river or lake, turning into tigers as they do so. Each evening they return home, reversing the process. The main settlements are Kandang Balok, ruled by King Uban, in Ulu Kemaman in Terengganu; another, ruled by King Paroi, on Mount Angsi in Negeri Sembilan; and Gunong Ledang, ruled by a king of the same name, on Mount Ophir in Johor. Gunong Ledang and his warriors were once beaten single-handedly in battle by Paroi, whose subjects, when not farming, read the Koran.[82]

Others say only the kings of such settlements, who are hantu belians, take human form inside the walls, their tiger subjects becoming men on leaving. These may have been the same tigermen, or weretigers, that, according to a report cited by Boomgaard, Chinese visitors to Malacca in the fifteenth century said were occasionally caught and killed on being recognized mingling with the crowds in markets. According to another report he cited—one by German physician Otto Mohnike, who visited the area in the 1840s—similar creatures were said to live in a village on Mount Dempo in South Sumatra. They too turned into men on leaving, visited markets, mixed with people, and even took human wives.[83]

Some people say *all* the tigers in these settlements are hantu belians.

One story, cited by Skeat and Wessing, has it that a shaman once invoked a hantu belian, who hauled him off to such a settlement when he did not pay the right fee. There he married a tiger princess, and when he was finally allowed to visit home he, too, turned into a tiger as he passed through or under the water.[84]

The Batek told Endicott *their* deceased shamans live in caves around a stone pillar, Batu Balok, on the Palah River: a pillar that is the center of the world and the home of Rajah Yah, king of the tigers. They assume human form on entering the caves, and tiger form on leaving. (Some say Batu Balok is also the home of Gobar.) Naga tiger-form ancestors also live in villages, but in tiger form. Ao Nagas, reported James Mills in 1926, say these familiars have a king who lives on Piyongkong, a mountain in the Phom area. At certain times of the year the familiars gather there, dance around their king, and throw food into his mouth. If the king is displeased with any of them, he eats the offender. And in Sunda, reported Wessing, bobongkongs are said to complete their transformation into tigers in a tiger kingdom on Mount Kandana.[85]

All the people of Daya, as well as becoming tigers post-mortem, are said to be weretigers too, noted Wessing. Meanwhile, Bakels was struck by how often the Kerinci attribute tiger-form ancestor characteristics to real tigers, how in stories the two types of tiger often merge.[86]

Indeed, so intimately is the connection made between people and tigers in Asia that the distinction between "real" tigers, "follow tigers," "village tigers," tiger-form ancestors of one sort or another, and even "tigermen," or weretigers, is often so blurred as to be almost meaningless. One final story, heard by Wessing in West Java, illustrates this well. A childless old woman called Emah Cowet was out in the forest one evening when she met a mother with a sick little girl. The mother pleaded with her to help the child, at which Emah prayed out loud, "I witness that there is no god but Allah and that Muhammad is his prophet," and blew on the child's head. A few evenings later a tiger approached Emah, told her not to be afraid, and said, "I am that small child whom you doctored that night and I would like you to adopt me." Emah consulted her husband, who agreed, at which the tiger said, "Thank you, mother. If ever there is anything, or you need help, just call my name."[87]

3

Killers and Killed: Propitiation and Appeasement

> *"Appeasers believe that if you keep on throwing steaks to a tiger, the tiger will turn vegetarian."*—Attributed to Heywood C. Broun (1888–1939)

As in Europe, people in Asia have long believed that when someone is denied proper funeral rites, or dies a violent or otherwise unnatural, premature death—including suicides, women who die in childbirth, and victims of such calamities as disease, attacks by wild animals, lightning strikes, and earthquakes—then until it is laid to rest their spirit restlessly haunts its surviving relatives. After the 2004 tsunami, for example, there were widespread reports in Thailand of sightings of victims' ghosts.[1]

In much of India, such bhuts or nats are said to be ruled over by Shiva, though they are often linked with other malevolent deities too. High-caste Hindus generally consider the bhuts of low-caste Hindus and Untouchables (or Dalits or Harijans as they are also known)—a group that includes all tribespeople—to be the most malevolent. Even those of low caste reckon that the higher your caste, the more likely a bhut is to molest you. Because of this, noted Elwin, Gonds chose names implying they were of even lower status than they were: like Bhangi, meaning sweeper, or even Ghurha, meaning dung heap.[2]

When someone in colonial India was killed by a tiger or a leopard, tribespeople, other Untouchables, low-caste Hindus, and poor Muslims alike believed that other members of the victim's community, especially their extended family, were then themselves in great danger of being targeted, because, it was said, the victim's bhut guided the animal to kill them too. The idea was based on bitter experience. Webber heard that the tigress that plagued the Juli road in Naini Tal in the 1860s took all four sons of a widow, while on the Moran River in Hoshangabad in 1867 Hicks bagged a tigress whose victims included three Gond brothers.[3]

One way a victim's bhut supposedly guided its killer, in the case of someone killed by a tiger, was by possessing a jackal, which then accompanied the tiger everywhere it went. Baker noted that in Bengal such a creature was called a *phuao*, *pheeow*, or *pheal*, in imitation of a jackal's cry, whereas in 1860 Indian Army officer Henry Shakespear wrote that such a creature was known as the Kola Buloo, "the tiger's provider." The idea seems to have arisen from sightings of jackals hanging around tigers, ready to scavenge their kills, and from imagining the truly dreadful cries that jackals utter, especially at night, as the wailings of lost souls.[4]

Another idea widespread in India was that the victim's bhut, which was thumb-sized, rode on the killer's head — or ran just ahead of it — steering it away from danger, urging it to kill again, and guiding it to its next victim. And as the number of victims and hence spirit guides grew, so the killer grew ever more cunning and dangerous. It was said, wrote Crooke, that the weight of its victims' spirits eventually caused its head to be permanently bowed.[5]

Russell and Hira Lal wrote that the Halba, a caste of cultivators and farm servants in central India, said the bhut lured victims within reach by calling out and offering them tobacco, and that in certain Gond communities, when a woman widowed by a tiger remarried, she went through the ceremony with a dog or ax, so the jealous and vengeful bhut of her first husband would guide the tiger to attack *that* instead of the groom.[6]

Much the same belief was found among the Batak, noted Wessing, and among the Muong (Figure 5) and ordinary Vietnamese, who, reported Baudesson, said the spirits were life-sized and rode on the tiger's back. When the Vietnamese set a pit-trap for a man-eater, they sprinkled corn all around, so the spirits would smell it and jump clear just in time. According to Baze, the Muong thought the spirits of a man-eater's victims powerful enough to upset a hunter's aim.[7]

Baudesson once led some Muong coolies on a reconnoiter of an area of jungle that a colleague, Sergeant Valutioni, had warned him was infested with man-eaters. He reluctantly allowed the headman's eager 12-year-old nephew Sao to join the expedition, and though sceptical about the sergeant's caution, kept the boy in front of him so he could cover him with his rifle. As Sao hacked his way through the foliage he hummed a plaintive tune in honor of Duc Thay. The heat was intense, but they kept going, intent on reaching a cool river by nightfall. The hour was late when

3. Killers and Killed

Figure 5. This photograph of Muong villagers on the Lagna River was taken on American traveler Henry C. Flower's excursion into French Indo-China in 1920. Some Muong once told William Baze, a farmer in upland French Indo-China before World War II, that man-eating tigers kept a tally of their victims by nicking one of their own ears with a claw each time they claimed one (*Asia*, October 1920).

they finally made it, and on the sergeant's urging the coolies lit fires and Baudesson gave the order that no one was to go down to the river without an armed escort.

But the coolies were so excited that Baudesson immediately relented and joined them all in a rush to the water. By the time they had finished bathing and filled their gourds, night had fallen. The way back was along a narrow path through dense bamboo. The coolies passed Baudesson in single file, Sao bringing up the rear and humming the same plaintive tune as before. "He looked so happy," recalled the Frenchman, "that I could not resist giving him a friendly pat on the cheek as he went by." Moments later came a heart-rending scream. Baudesson glimpsed the lad dangling from a tiger's jaws and automatically raised his rifle, but before he could take aim, man-eater and victim were gone. "The Lord Tiger! The Lord Tiger!" cried the coolies. Dazed and distraught, raging at their impotence, Baudesson and the coolies returned to camp.

There Baudesson tried to console Sao's uncle, who, through his tears, said that the same fate had befallen Sao's parents, that their restless spirits had "long demanded another companion." Baudesson then heard murmurings among the coolies: the white men had angered the spirits and brought this evil upon them, they said, and they would be deserting at dawn. Backed by Sao's uncle, the Frenchman nipped the rebellion in the bud by demanding that the coolies hand over their ID cards, which bore each man's fingerprints. There was no danger of them fleeing now, for without his card no coolie could find work. But no one slept that night. Wrote Baudesson: "The dog trembled and whined as if scenting evil. The tiger must have been watching us!"

The next morning the coolies found poor Sao's remains, wrapped them in palm leaves, and dug a grave where they lay. Sao's uncle asked Baudesson for a piece of paper, on which he sketched two adults and a child riding a tiger. There had not been room on its back for another adult, he said, so the tiger had targeted Sao and not Baudesson: normally, tigers much preferred the flesh of white men. The headman burnt the drawing and scattered the ashes over Sao's shrouded remains while murmuring prayers. Then, after filling in the grave, the diggers strode around, crying out to "the High and Mighty One" to seek no more victims.[8]

Anxious to restore confidence among the coolies, Baudesson declared his intention to destroy the man-eater. Aghast, the men did their best to dissuade him. "They feared the vengeance of the tiger," he wrote, "but I was not to be turned from my purpose." Fourteen nights running he sat up over a tethered deer, but the tiger never came; though they nightly heard it roaring nearby. Finally he acceded to the coolies' desire to strike camp. A few months later the tiger devoured a fellow missionary, Lieutenant Gautier, at the same spot.[9]

In China — and, as Locke noted, among Chinese coolies abroad — the spirit of someone killed by a tiger was called a *ch'ang kwei*, which de Groot translated literally as "the ghost of him who lies flat upon the ground." Some say such a ghost, which apparently is the size of a small child, lacks the courage to go elsewhere. Others say the tiger enslaves it, even compelling it to reanimate the corpse and make it undress — even fold the clothes up into a neat pile — to spare the killer the inconvenience of getting cloth stuck between its teeth. A ch'ang kwei guides the tiger to the next victim in order to free itself from the animal's service by obtaining a substitute. In the meantime it does everything it can to protect its charge.[10]

One story recorded by de Groot tells how a professional tiger-slayer once kept finding his crossbow-trap sprung, with paw prints all around, and the arrow loosed, but no sign of the tiger being hit, so he hid in a tree and kept watch. Before long he saw "a little sprite in blue garments, with hair growing to a level with its eyebrows" come along ahead of a tiger and harmlessly trigger the trap. The hunter nipped down and fixed another arrow, but the same thing happened. So he tried again, and this time the tiger came back ahead of the sprite, and was killed. At this, the sprite hopped about excitedly, then vanished.[11]

A few ch'ang kwei are brave enough to lead their tiger-masters into traps. Thus one day in A.D. 755, reported de Groot, a boy saw a ch'ang kwei walking ahead of a tiger, and when this happened again and again he realized he was next to be killed. He calmly told his parents he would then lead the tiger through the village, where they should prepare a pit-trap in readiness. Sure enough, shortly after that a tiger killed him. He then appeared to his father in a dream and said: "I am a ch'ang kwei now; tomorrow I shall take the tiger thither; be quick to prepare a pitfall on the west side." His father did so, and bagged the tiger.[12]

Similarly, in 1915, folklorist Frank Hamel wrote that there was a saying in the Sagar and Narmada territories of central India — which included the Narsinghpur, Damoh, and Jabalpur districts — that once a tiger killed someone the victim's spirit stopped it killing again, while Perry cited a sportsman in India who was sitting up one night for a tiger that had killed someone when the victim's spirit tapped him on the shoulder to warn him the animal was stalking him. And according to Bangalore-born and bred sportsman Kenneth Anderson, writing in 1959, the Chenchu tribespeople of southern India believed a victim's spirit would not only *expect* surviving relatives to exact revenge on the animal, but would be outraged if they accepted any reward for doing so.[13]

Like the Muong, people of all religions in colonial India reckoned the spirit of someone killed by a tiger or leopard could never be laid to rest until at least some of the remains, however small, had been recovered for cremation or burial. Muslims were also anxious to give the remains their customary swift internment. But by law, noted Perry, to prevent people murdering their enemies then claiming an animal had killed them, all such deaths had to be reported to the authorities and the remains left in place until the police had inspected them.[14]

Yet even if this was done — and it is hard to believe it always was — sportsmen would often urge the relatives to leave the remains in place overnight, so they could sit over them and try to bag the killer. Alternatively, as in the case of one of several man-eaters Anderson bagged in the 1940s, '50s and '60s, they would simply keep quiet about finding any remains until the following day.[15]

Not all colonials were so cavalier. In the Raipur district of central India in 1856—only a year before the Mutiny—Shakespear tried in vain to persuade his Muslim troopers to leave out the remains of one of their comrades who had fallen victim to a man-eater the sportsman was hunting. As he recalled: "I talked a good deal to the Mussulmans [Muslims] about our both being men of the book, and not infidels; that they were of the same opinion as I was, that when the soul had fled, the remainder was but dust; that I would just as soon be eaten by tigers or jackals as be put into the finest mausoleum.... But they thought differently, and took away and buried the body."[16]

In the Malay Peninsula, remains had to examined at a hospital for official confirmation of cause of death before they could be interred, and in Terengganu Locke was initially thwarted in trying to kill a man-eater in the area of Jerangau in 1950 and 1951 by Muslim relatives immediately removing remains so the examination could be done as soon as possible. He even made a dummy corpse out of scraps of the fifth victim's bloodstained clothing in an effort to bag the killer. When that failed, he wrote to all the local headmen saying that the next person killed was not to be moved under any circumstances. As he explained: "There was not the slightest intention on my part of allowing the man-eater to feed from the body. It was essential, however, that the corpse should be permitted to remain as an attraction to the man-eater to return to that spot." When this was finally done, with the eighth victim—more through fear of the tiger than a wish to comply with his instructions, he noted—Locke was finally able to shoot the tiger. But a Malay official complained to the state government that he had used a dead Malay "as bait, in order to enjoy the sport of tiger shooting." Locke explained to the head of the state religious department that he only ever shot livestock-lifters and man-eaters, and that he greatly respected Malay beliefs, but said that it was surely better to leave a body out for a few hours than allow a man-eater to kill again. Finally the mufti of Terengganu, the Muslim legal expert who had the final say, ruled that what Locke had done was "a praiseworthy undertaking and one that is esteemed under Islamic law."

One curious aspect of this case was that after the man-eater claimed its sixth victim two Malay pawangs independently told Locke it would be impossible for him to shoot the tiger until it had killed its eighth person: "If he does not kill you first, *Tuan*," one added.[17]

In Burma, Christopher noted, Karens, Shans, and Burmans alike said that by deliberately mutilating remains you could so anger the victim's spirit, which rode on the tiger as in French Indo-China, that it would steer the animal straight back to them, where it could then be shot. But it took a brave man to do such a thing, for while some people were prepared to tackle man-eaters, many others—as Baudesson found with the Muong— were afraid even to aid such an enterprise for fear of inciting not only the victim's spirit but the tiger itself, or the deity behind it, to target them in revenge if the attempt failed. Fayrer even cited a case in Chota Nagpur of some "natives" who, after helping a sportsman kill a man-eater, ran away and hid for several days, because they believed "it would be revenged on them, and take the form of a human corpse so as to get them hanged for murder." In Burma, wrote Christopher, just carrying a weapon—even a stick—was said to so anger a man-eater that it would pick the offender out from a group walking "Indian file" rather than take the last man in the line—like Sao—as was usual.[18]

In 1933, sportsman Major Leonard Handley recalled hunting a tiger that killed mainly Gond women collecting firewood and thatch in the Danauli Reserved Forest near Amarkantak. One day three Gonds had just bravely recovered a woman's corpse—which was minus the left leg— when, emerging from the trees, they bumped into the local forest guard who, acting under Handley's orders, told them to put it back where they had found it. Unwilling to risk their lives again, they instead left it in the fork of a nearby tree and lit fires underneath; presumably to keep the man-eater away. The next morning, when Handley saw this, he ordered her four strapping young sons to get her down. To his disgust, only by losing his temper could he get them even to touch her, and after letting her fall to the ground they threw her into a makeshift grave "like a dog."[19]

Handley attributed their reluctance to touch their mother to fear of the evil spirits they believed had entered her body after a night in a tree. Many Gonds and others, like peninsular Malays, who lived outside the forest, even if on its fringes, did indeed fear it, imagining it to be full of evil spirits, unlike those tribespeople who lived inside it, people to whom it was home, a place of cool sanctuary. But there is a another, more likely explanation, for as Russell and Hira Lal noted, Gonds and Korkus alike were reluctant to touch anyone who had merely been mauled by a tiger, never mind killed by one, lest they become the animal's next target. Fuchs wrote that Gonds and Baigas were even careful not to walk on the blood-soaked ground where a tiger had killed someone, for fear of attracting the animal's attention. J. B. H. Thurston similarly reported that Chinese coolies in the Malay Peninsula would refuse to touch the body of anyone killed by a tiger.[20]

In 1956, long-time Seoul resident Mary Linley Taylor wrote that Korean villagers usually tried to keep a man-eater's kills secret for fear of "offending the tiger," while Bradley heard at Fort de Kock (now Bukit Tinggi) in West Sumatra that villagers there would tell no one about a man-eater for fear that, if it was killed as a result, its spirit would "revenge itself upon the informant, or send other tigers to do it."[21]

Villagers in India — especially those of tribal origin — would likewise often keep quiet about a man-eater in an attempt to earn its gratitude, but more than that would try to propitiate or appease it with prayers and offerings.[22]

Around 1900 in Chanda an old Gond told forest officer D. King Martin the following story, which he swore was true. When he was a young man, he said, a tigress "haunted" his village and two neighboring villages for months on end, claiming a mounting toll of lives, before one day vanishing into thin air. After several weeks with no sign of her the villagers concluded she was a malevolent goddess who had finally had her fill of them; so to keep her sweet they began to worship her with animal sacrifices and other offerings. When several months passed uneventfully the men of the villages—forestry coolies—at last decided it was safe to celebrate their liberation from tyranny, to which end they held a party one evening at their communal forest hut. Much feasting and merrymaking ensued before eventually they settled down to sleep: and sleep came quickly, for as the old man admitted they were all slightly the worse for drink. Two brothers, Sapru and Dumru, who only that morning had sacrificed a final goat to the goddess, lay side by side.

That night, said the old Gond, the goddess reappeared and entered the hut in search of one last victim: "She paused by the brothers and her hot breath played about them; her eyes sent out their message and Dumru arose and followed her. Sapru, sleeping also, arose and followed at a distance, drawn by the power of her eyes to witness the sacrifice." The tigress led them to a dry river-bed where her three cubs were waiting, before turning and pouncing on Dumru. Waking from his trance, Sapru could only look on as there, on a carpet of silvery sand in the brilliant moonlight, she toyed with his brother like a cat with a mouse, while her cubs watched, absorbing the lesson. At last they killed and ate him, leaving Sapru, frozen with fear, to live to tell the tale. And from that day on, concluded the old Gond, the goddess never troubled them again.[23]

With all this, many colonials were struck by how indifferent — blasé, even — to their own individual fate were people who lived alongside tigers.

Baze recalled that when camping in the jungle he routinely ordered his tribal followers to keep a protective ring of fires burning throughout the night, and that all the time he was awake they would do so, but the moment he dozed off, so would they. When admonished, they would say, "Well, if a tiger had come he would only have taken *one* of us!" Likewise, they were happy to stride along in a line — like Sao's fellow Muong — just so long as *they* were not the one bringing up the rear.[24]

If you were predestined to fall victim to a tiger, well then, there was nothing you could do about it. Best recalled that a party of woodcutters was once working away in a known haunt of the notorious man-eater that was so fond of eating mail runners in Balaghat, when one of them spotted the tiger approaching and raised the alarm. Each man shot up the nearest tree and waited for the animal to move off. But instead of leaving, the tiger stopped under one particular tree and settled down to wait, evidently having chosen that tree's occupant as its next meal. Several hours passed and still the man-eater did not budge. Eventually the trapped man shouted to his friends to go home and get help. They needed no second invitation, and were off in an instant. Back at the village, however, they decided to a man that a good meal and a long snooze were in order, so it was several hours before they returned bringing help. With the tiger driven away at last, their mightily relieved companion was finally able to climb down to the ground. As he merrily made his way home with the others, the tiger sprang out and carried him off. (Best read about this case in the *Central Provinces Gazette*, and wondered if the man was much plumper than his companions, and so a more tempting target. Another explanation is that he was singled out for having transgressed.)[25]

In Java, Wessing heard that those born at sunset in the first month of the Javanese year were fated to be killed by a tiger one sunset in the same month, no matter how careful they were. In Sumatra, he notes, the Batak believe your soul chooses the manner of your death before you are even born; so again, you cannot avoid your fate. A 1916 report told of a Batak man who fretted constantly because he was told his soul had chosen for him to be killed by a tiger: until one day a wooden carving of a tiger fell on him and crushed him to death.[26]

Faced with creatures they believed were not only agents or manifestations of angry deities but were guided by their victims' spirits, most shikaris were understandably nervous about sitting up over human remains. When in 1861 Forsyth spent a month's leave shooting in the hills surrounding Jabalpur, he did so with an old shikari called Bamanjee as his guide. One evening they sat up in a machan over the carcass of one of Forsyth's baggage ponies that a man-eating leopard had killed. Knowing the leopard was unlikely to show until well after dark, if at all, Forsyth initiated a whispered conversation to pass the time. Had Bamanjee ever sat up over human remains, he wondered? Yes, replied the old man, but he did not fancy it much, as the body stank abominably: and that was not all.

The nearby village of Ponhri, he began after much pressing, was once haunted by a perfect *shaitan*, or demon, of a tiger; a true monster. It lived on the hill between the two paths up to the village, and whenever the crafty old creature saw a party leave Ponhri it would dash over to whichever path they took and wait in ambush before springing out with a roar and carrying one of them off in a flash. Likewise the villagers saw it pounce on many an approaching traveler. Sometimes they had time to warn the man to take to a tree, but usually the tiger was too cunning for them, stalking its victim with extraordinary stealth.

In Ponhri, continued the shikari, there was a Gond thakur called Padam Singh, a man who had once shot a tiger and so was considered an authority on such matters. When one day the tiger was so bold as to enter a cowherd's hut and pounce on the unfortunate man when he came home for his dinner, Padam Singh proposed an expedition to follow the drag and rescue his remains. Taking whatever weapons and loud instruments they could lay their hands on, the men of the village all followed him, and at their noisy approach the man-eater abandoned its meal. Seeing the corpse was only half-eaten, the upper half remaining untouched, Padam Singh — who owned the only matchlock in Ponhri — suggested he sit over the remains in a tree and await the tiger's return. The relatives reluctantly consented, and the brave Gond was left to his vigil.

The man-eater soon returned, but before it was within range it paused; and to Padam Singh's horror the legless corpse raised its right arm and pointed straight at him, at which the tiger fled. A short while later it came back, but the corpse warned it away again in the same spooky manner.

Recovering his composure, Padam Singh had an idea. He whittled two sticks, climbed down, and thrust the stakes through the dead man's hands, pegging them fast. He barely had time to climb back up the tree before the tiger returned. Again it paused, but on receiving no warning this time came forward and sank its teeth into the neck of the corpse, giving the Gond an easy shot.[27]

Bamanjee's story — with its notable similarity to de Groot's crossbow-trap yarn — is an archetypal one in India, various versions of it having been recorded over the years. Burton told one in 1931, and Powell another in 1957. In the latter, three shikaris sitting over a man-eater's kill one night make the fatal mistake of only pegging down one arm. When the killer returns, the corpse sits up and points at them again with its other arm, at which the tiger — a "somewhat supernatural" one, noted Powell — springs the thirty feet (no less) up to the machan and slaughters all three men.[28]

Some shikaris, it seems, were so scared of sitting up for man-eaters that they encouraged people to think such creatures were supernatural to save face when they declined the task. Take the following case reported by Maurice Hanley, writing as Patrick Hanley. With the usual "fiendish cunning" a leopard killed one man, eight women, and three children in a small village near Kumbha in Borpatra in the Naga Hills in the space of five months. It snatched its first eleven victims when they ventured outside after dark to fetch firewood or draw water from the village well. Then, when the terrified villagers took to barricading themselves in at night, it resorted to hurling itself against their flimsy bamboo doors. Twice it gained entry this way and mauled four people, one fatally, before the occupants bravely beat it off.

Despite the government's offer of the maximum 500-rupee reward for its destruction, the village shikari — whose own wife was among its victims — was too scared to tackle it, and instead spread the story that it was a demon, a ghost leopard that no bullet could kill, simply to cover up for his cowardice; and to prevent this getting out he repeatedly talked the village headman out of asking any sahib from the nearby tea plantations to tackle the animal.

One day the shikari went out to shoot some deer he had seen eating the villagers' rice crop at the edge of the jungle. Chancing on a leopard lying fast asleep in a rocky hollow, he shot it dead, and wheeled it in a cart to Hanley, who gave him 20 rupees for it, for it had quite the best pelt of any leopard the planter had ever seen. With the shikari looking on, Hanley started to skin the animal. As he did so he noticed that the joint of its right hindleg was greatly enlarged. Opening this up he found a rough lead pellet in the smashed bone. At this, he declared that the shikari had unwittingly killed the man-eater.

On hearing this, the man first trembled and sweated at the thought of what might have happened had he woken the leopard, but to the planter's amusement he then began boasting of his great courage and fine marksmanship. When he started talking about himself as a hero and the savior of his people, however, Hanley could stand no more. The pellet could only have been fired by a muzzle-loading rifle, as used by shikaris. Most likely, then, Hanley admonished the man, he had created the man-eater himself through negligence and cowardice in not following up and killing the leopard after wounding it.

Now feeling rather sorry for himself, the shikari admitted that five or six months earlier he had indeed shot at a leopard that had got away. He begged Hanley not to tell his fellow villagers, for if he did they would surely drive him out of the village. Hanley generously obliged. He later even obtained the 500-rupee reward for him from the district commissioner.

The village headman, Hanley noted, claimed his own share of the credit for the leopard's demise on the basis that he had had the foresight to make an offering to the jungle gods the night before.[29]

When a tiger or leopard killed a villager in colonial India, the relatives would typically pay a tribal priest to perform various complicated magic rites—none of which came cheap—to exorcize the animal, lay the victim's spirit to rest, and appease the offended deity. (All over India, indeed, such priests performed similar rites for relatives of people who died from other sudden, violent or otherwise "unnatural" causes, like snake bite and cholera.)

Some priests—generally from the remotest, smallest tribes—were said, not least by themselves, to be especially powerful. In the Terai there were three such tribes—the Tharus, Bhoksas, and Mechas—who mixed freely with each other but with no outsiders bar Banjaras. And like the Banjaras, all three venerated Bhawani and Bhairava. Tharus took part in mass animal sacrifices to Bhawani at Devi Patan in the Gonda district. Like many shamans in the Malay world, Bhoksa and Tharu priests openly claimed the power to control evil spirits and wild animals: a power, noted Eardley-Wilmot, that in the case of Tharu priests was "accompanied by the fleshly lusts of gluttony and drunkenness."

In a reminiscence that was surely met with raised eyebrows by most fellow colonials, Eardley-Wilmot recalled stopping sometime in the 1870s with a fellow forest officer named Dodsworth at a Tharu village just inside Nepal, where the Mohan River forms the border with India, and asking the inhabitants if they knew of any tigers they might shoot. Not knowing the sahibs, the Tharus were initially coy, but finally they introduced them to their local priest, who they paid to keep tigers away from their cattle.

The gift of a goat and a bottle of rum plus the promise of a large reward soon won this man over, and the next morning, riding the largest of the two jungle-wallahs' five elephants, murmuring incantations, and tinkling a bell, he led the sahibs out, and stopped at one of the least promising places imaginable, a patch of bare grass. Sure enough, their first sweep drew a blank. But on the second sweep Eardley-Wilmot's mahout spotted a crouching tiger, which instead of fleeing or charging got slowly to its feet, walked out of the grass, turned broadside on and stopped, presenting Eardley-Wilmot with the easiest of shots.

The two sportsmen begged the priest to repeat the experiment, and moving on they bagged another tiger with equal ease. Wrote Eardley-Wilmot: "Bewildered, we returned to camp. I forget at this distance of time the number of tigers this peculiar man presented to us in this manner; they were but few, as he explained that if there were no tigers there would be no fees for preserving the cattle, but they were sufficient to fill us with astonishment, and also with faith in his power."

Back at their bungalow at Dudhwa, just south of the river, the two sportsmen decided to try the man's methods for themselves on an old tiger of the vicinity that had long evaded them by its "cunning." To this end they tried an incantation Eardley-Wilmot's orderly happened to know, while ritually lighting little oil lamps and laying out offerings of rice and various spices. The result was that the tiger strolled in front of their elephants as if dazed, but despite firing repeatedly, neither Eardley-Wilmot nor Dodsworth, the finest shot Eardley-Wilmot ever knew, could hit it. They concluded that the charm had been potent enough to call up the tiger, but not powerful enough to ensure they hit it.[30]

Tribal priests all over India laid claim to similar powers. In Mysore, according to Campbell, a small hill tribe called the Eriligarus were said to be so good at charming tigers that their women entrusted their children to the animals' care when in the jungle. Abul Fazl Allámi's sixteenth-century *Aín I Akbari*, which documents the administration of the great Mogul emperor Akbar's empire, says of Gonds, Bhils, and Kols alike that they "can tame lions, so that they will obey their commands, and strange tales are told of them," while Burton recalled his followers in the Deccan saying that Gonds protected tigers by deflecting sportsmen's bullets. But Gonds themselves and other tribes and castes in central India bowed to Baigas, who had lived in the forest the longest and were therefore considered the real masters. Elwin recalled how on a rainy May 18, 1933, a policeman relieved the tedium of the crowd in the pastoral officer's dispensary with

Tracking the Weretiger

stories about the incredible powers of Baigas. A Baiga priest can shut a tiger's jaws and lead it about by one ear, he declared, adding that when Lord Reading, Viceroy of India (Figure 6), came to Mandla in the early 1920s, a Baiga priest warned all the tigers to leave the jungle, and the viceroy had to pay him 100 rupees to call them all back. The audience lapped up these stories, noted Elwin, especially the Baigas.[31]

Crooke recorded the following charm used by Baiga priests to render tigers powerless:

> Bind the tiger, bind the tigress,
> bind her seven cubs!
> Bind the road, the footpath,
> the fields through which they wander!
> Aid us, O Lord Krishna, and Lona Chamarin [a noted witch]!

To make it work the priest had to bathe seven times on seven Tuesdays. Crooke thought its efficacy doubtful. He also recalled once asking a colo-

Figure 6. Lord Reading (center), Viceroy of India from 1921–1926, with the result of a typical large-scale tiger beat in Nepal. When he went on a similar hunt in the Mandla district of central India, reported local pastoral officer Verrier Elwin, it was wildly rumored that a Baiga priest in his pay led thirty-two spellbound tigers one by one before his elephant for him to shoot.

nial official in the Mirzapur district of northern India about the supposed power of Chero tribal priests to control tigers there. The man replied that when he first arrived and a Chero priest offered to protect him from the many tigers in the area, he told him that he could look after himself, and advised him to do the same. That same night the priest was eaten by a tiger.[32]

In his official 1867 report, as cited by Russell and Hira Lal, Hoshangabad Settlement Officer Charles Elliott wrote that when a tiger encroached on a Korku village there the village priest would confidently make an offering to the Korku tiger god to keep it away, believing the tiger "never fails to fulfil the contract thus silently made, for he is pre-eminently an honourable upright beast, not faithless and treacherous like the leopard whom no contract can bind." It was said that one priest had a saj tree into which he would drive an iron nail, after which the tiger would ratify the contract by scoring the bark with its claws, and that some priests could order two or three tigers at a time to cower at their feet. But some priests, reported Elliott, lost their lives to tigers, having misplaced confidence in their powers. Russell and Hira Lal noted that belief in the powers of Korku priests had declined somewhat since Elliott's day.[33]

In the song cited by Banner, the Javanese flattered the tiger that it was stronger than iron, a well-known scarer of evil spirits worldwide. Russell and Hira Lal noted that in the temples where Gonds once sometimes sacrificed people to Kali, an iron plate over her effigy's face protected worshippers from her wrath, and throughout India tribal priests hammered iron nails into door and bedposts, as well as trees, to keep out evil spirits. Fuchs reported that the origin of the Kusru Gond clan's special relationship to Bagheshwar was said to be that there was once a man-eater against which the most powerful spells were of no avail: until a sorcerer of the clan went into the jungle one night, cornered the tiger, and successfully exorcized it by driving an iron nail into a tree. Fuchs also learned that iron nails featured prominently in the rites performed by select, mainly Baiga priests whenever a tiger killed a Gond or Baiga during his time in Mandla.[34]

Following an attack, one priest told Fuchs, the victim's closest kinsman recovered any remains then took a few rupees and a bottle of mahua spirit to a qualified priest like himself. After drinking the whole bottle the priest went off into the jungle to dig up a secret magic root, while the kinsman hurried home to help prepare for the cremation of the remains, which had to be done quickly before the tiger, watching close by, could reclaim them.

Escorted by the priest, the victim's kinsmen carried the remains on a bier to a burning-ghat — a riverside cremation site — with the other men of the village walking not behind, as was usual, but on either side, to shield the remains from the lurking tiger. Without ceremony a pyre was heaped up around the remains, and the closest kinsman lit it at either end. Instead of then leaving, as was normal, everyone surrounded the pyre in a protective ring until the flames met in the middle: otherwise, the moment they left, the tiger would dive into the river to wet its coat, leap onto the pyre, and reclaim the remains.

All then bathed in the river bar the priest, who buried a piece of root at either end of the ashes and drizzled mahua spirit on them, reciting incantations all the while, before eating some root and finishing off the bottle.

Meanwhile the women spring-cleaned the victim's hut and gave it a fresh coat of cow dung, the dwelling having been tainted by its owner's sudden death.

Back at the village the priest buried a piece of root outside the victim's hut and hid a sliver of the same in a crack in the door or wall inside, where he drank another bottle of mahua spirit, while the men shared a bottle of their own and discussed the serious matter of how much the closest kinsman would have to pay for the main rites, which were yet to come: the *Matti uthana*, or "lifting of the soil," to exorcize the tiger; and the *Thor bel*, or "breaking the string," to appease the deity controlling the tiger, and keep the tiger away for good.

The priest and the men negotiated the fees, which the former initially always set unrealistically high, but which ultimately depended on the closest kinsman's means. For the Matti uthana even a poor man had to provide food and mahua spirit for two feasts, plus a few rupees. The fee for the Thor bel was even higher: at least 50 rupees, plus all the provisions for a feast, and as much as a bullock or a buffalo if the man could afford it. The negotiations were long and drawn out, the closest kinsman plying the priest with mahua spirit as he strove to get him to lower his price, while the priest sat poker-faced, holding out for the highest amount: the villagers needed him, and he knew it. If he persisted in demanding too much, other kinsmen, fearing for their own lives, offered to contribute. The matter finally settled, some of the men escorted the priest home, plying him with even more mahua spirit until they almost had to carry him into his hut.

Early the next morning the somewhat worse-for-wear priest again went into the jungle to dig up more root then, accompanied by the men of the village, staggered to where the victim had been killed. There he drank some much-needed hair-of-the-dog, and chose a *barwa*—a man who falls readily into a trance—as an assistant. After sharing a hookah—a tobacco-pipe with water to cool the smoke—and yet more mahua spirit with this man he ordered the agreed food, mahua spirit, and cash payment to be left on the village boundary. That done, the Matti uthana could begin.

The men retreated to form a large ring around the site, and began making a racket. Meanwhile the priest began chanting, hammered an iron nail into the nearest tree, and held out a piece of root for the barwa to sniff. Eventually the barwa fell into a trance, "the spirit of the tiger" possessed him, and he began to growl, then roar, and prowl on all fours. At this the priest scraped up some of the blood-soaked soil and pressed it into the man's hands, at which the barwa bounded away on all fours, dashing across clearings, and slinking through the undergrowth, while the men gave chase, yelling "Kill him! Kill him!"

Once at the village boundary, the barwa halted under a tree, where the men surrounded him at a safe distance and waited for the priest, who had been following at his own leisurely pace. The priest sat by the barwa, who trembled in every limb, and commanded him to sit too. He held out a piece of root for him to sniff once more, and the barwa slowly came out of his trance, aided by some mahua spirit.

Chanting all the while, the priest laid seven iron nails on the ground and poured a handful of grain alongside them. As soon as a cockerel and hen provided by the villagers began pecking at the grain, he twisted their heads off and spilt their blood over the nails. The men lit a fire, and the priest poured ghee and powdered grass into the flames, gave a nail to the barwa to hammer into the tree, and told him to join the others: his job was done. He then poured mahua spirit over the remaining nails, and hammered two of them into the tree.

Finally the priest took the other nails to where the villagers drew water from the river—presumably because this was a likely spot for the tiger to lurk in ambush—hammered two of them into the ground, killed another cockerel, and spilt its blood plus some mahua spirit over them. With that, the Matti uthana was done.

The priest took the last two nails and the dead birds home with him, plus half the food, mahua spirit, and cash. The barwa took the other half, for by handling earth soiled with the victim's blood both men were temporarily put out of caste and would have to give a feast to gain readmission.

Different versions of this ceremony were recorded in colonial times. Russell and Hira Lal reported that in Mandla the priest made a cone out of the blood-soaked soil, representing either the victim or a relative, and seized it in his teeth like a tiger. The villagers then buried it in an anthill over which the priest sacrificed a pig. The next day they took a chicken to the site, where the priest marked its head with ochre, representing the victim, and threw it into the air saying, "Take this and go home." The purpose of the ceremony, reckoned the ethnologists, was both to lay the victim's spirit to rest and exorcize the tiger. (They further reported that Gonds reckoned it took ten to twelve days for the victim's spirit to come to rest, the relatives helping it along by winding a copper ring on a thread and suspending it over a pot of water, pouring the blood of sacrificed animals over it, and keeping watch until it finally fell into the pot.) Forsyth similarly reported that it was the priest himself who played the role of the tiger, and that he reenacted the killing by eating a mouthful of the blood-soaked earth, after making offerings to "the manes," which can mean either a person's ghost or the deified souls of ancestors. Whichever, if after eight days the priest himself had not been killed by the tiger, the victim's spirit was deemed to be at rest. Fuchs himself noted that the Matti uthana was also performed by various castes and tribes, including the Munda, in Chota Nagpur, where its purpose was to drive away not the tiger but the victim's spirit. But as Elwin for one observed, the purpose of the nails was clearly to shut the tiger's jaws and keep them shut; or at least keep the animal beyond the village boundary.[35]

Elliott may have reported that claw marks around a nail represented a tiger "ratifying the contract," but Russell and Hira Lal heard that a Baiga priest in the Banjar Reserved Forest of Mandla once so successfully shut a man-eater's jaws with a nail in a tree that claw marks showed where the tiger had tried to remove it and so regain its powers. A more mundane explanation is that the tiger was marking its territory, although in colonial times it was thought that tigers scratched trees to clean and sharpen their claws. Trees that tigers in India typically scratch include *Butea monosperma* (formerly *Butea frondosa*) — the kino, or bastard teak — and, appropriately enough, the "biga," or "bijasal" (*Pterocarpus marsupium*, also known, confusingly, as the kino, or bastard teak). Both ooze a red gum that people say tigers like to see flow like blood.[36]

In the 1890s, missionary Archibald McMillan heard in Balaghat that when a tiger killed a Gond or Baiga, fellow villagers sometimes hid some of the remains in a hole in a nearby tree. The tiger god then possessed the attendant Baiga priest, who pounced on a goat and tore its throat out with his bare teeth, like the celebrants at a Kusru Gond clan wedding. More commonly, though, the villagers gathered around the remains, threw a rupee on the ground, then tried to restrain the priest from picking the coin up in his teeth. If they failed, the tiger would kill more of their number, but if they succeeded it would move on and harass another village. McMillan's bungalow was in the Gond village of Khursipar, and when a tiger once killed a man from the neighboring Gond village of Karamsara, the local priest sacrificed a chicken to the tiger god in Karamsara but buried the bird in Khursipar, in an effort to shift the man-eater there. When a deputation from Khursipar complained to McMillan about this, he suggested they rebury the chicken outside his bungalow, since unlike them he owned a gun and could defend himself if the tiger came after him. But the villagers would not hear of it. Wrote the missionary: "They probably feared that if my offer were accepted, and I should afterwards be killed by a tiger, the 'powers that be' might hold the village responsible for my demise." (Forsyth wrote that during outbreaks of disease in Mandla, Gonds and Baigas had the "pleasant notion" that it must be transferred somewhere else, to which end they swept the village clean and dumped the trash in the road or even in another village.)[37]

Returning to Fuchs's account, after the Matti uthana the victim's closest kinsman called on the priest to fix the date of the Thor bel. If the family could afford it, this was the day of the full moon in the next March–April or December–January period, whichever came first. If not, the ceremony might be postponed for a year, or even longer, to allow them time to get the necessary funds together. But it could only go ahead if in the interim the Matti uthana proved successful: in other words, if no other relative was killed by a tiger. If one *was* killed, the Matti uthana had to be performed — and paid for — all over again. Clearly, then, when a persistent man-eater was at large a priest might be called on to perform one Matti uthana after another.

On the day of the Thor bel all the victim's family and friends gathered by the river, presumably at the same spot as before. They brought with them: the priest's fee; a goat, other food, and mahua spirit for him to hold another readmission feast; a sacrificial piglet, black hen, and black cockerel; and all the provisions for a feast for themselves. Before joining them, the priest hid another sliver of root in the door or wall of the victim's hut, hammered one nail into the door and another into the front gate, and poured mahua spirit over both.

Down at the river he had the victim's closest kinsmen wade into the water and form a circle, each man facing outwards, then wound lengths of string and weak rope around them all, each man holding them with clasped hands. The string was made of new, unbleached cotton, the rope from the fibrous bark of either a creeper called *dokar bel* or a tree called *kaboti*. Neither was ever used for anything else, because if any of the men ever touched either of them again in his day-to-day life, he too would be killed by a tiger.

The senior man of the circle held the ends of the rope and string, ensuring they overlapped. The priest poured a little grain into this man's cupped hands and held the cockerel and hen in front of them. As soon as they pecked at the grain he twisted their heads off and spilt their blood into the river. After pouring mahua spirit on the rope and string he cut them on either side of each man, chanting all the while. Each man passed the cut pieces under his right thigh to the priest, who wound them into a ball and stuffed them under a boulder in the river with the sacrificed birds. Finally the priest hammered a nail into the boulder, killed the piglet, and spilt its blood over it. With that, the Thor bel was complete.

Back on dry land the priest lay on a sheet onto which the men threw his fee coins as they stepped over him in turn. He walked home with his fee and provisions without looking back, leaving them to hold their own feast there and then or back at the victim's hut.

Just as there is uncertainty as to the precise purpose of the Matti uthana, so there is uncertainty as to that of the Thor bel. Fuchs wrote that it was meant to protect the victim's relatives for all time from the deity that sent the tiger, then added that another explanation was that it was meant to exorcize the victim's spirit. Whatever, he was assured that when both ceremonies were properly performed the tiger always left the area. This might have been true when someone was accidentally killed by an ordinary tiger, but with persistent man-eaters, besides being a community-affirming tradition and offering some sort of closure, repeated Matti uthanas or Thor bels arguably achieved little other than to line the pockets of priests, even allowing for their caste readmission expenses. McMillan recalled some villagers in Balaghat who had long suffered the depredations of a man-eater saying to him: "We cannot understand it! During the last four years, we have had man after man killed, so have changed our pujári [priest] no less than nine times. The most skilful and best-known within a day's journey have been here, and have done their best, yet they are all unable to stop these ravages!" McMillan noted that while such experiences eroded tribal people's faith in the ability of their priests to protect them from tigers, they would never countenance the idea that their tiger god did not exist, or that tigers had no supernatural powers.[38]

Tribal priests all over India performed similar rites. Russell and Hira Lal wrote that in central India priests of the Bhatra tribe demanded not only a 200-pound sack of rice, but all the victim's jewelry—just in case the victim's spirit should ride the tiger into the village in search of both— while priests of the Dumal agricultural caste demanded a share of the victim's property equal to what he would receive were he one of the family.[39]

As Captain R. H. Gribble of the Burma Frontier Service observed in 1944, the result of just the everyday animal sacrifices required to propitiate their myriad deities—never mind those special ones demanded in times of crisis—continually deprived poor villagers in India of precious livestock, perpetuating their poverty. It was in village priests' own interests to maintain this state of affairs not only for personal gain but to retain their power and prestige. In Fuchs's day the new independent Indian government rarely granted tribal shikaris permission to own guns, fearing they would use them to poach game, so tribespeople had little choice but to tackle man-eaters with magic. But in earlier times the seemingly supernatural powers of the worst man-eaters often deterred even the bravest armed shikaris. Small wonder, then, that sahibs like Forsyth and Hicks saw themselves as saviors whenever a man-eater terrorized a community.[40]

4
Fighting Back: Jungle-Wallahs to the Rescue

"A man-eater is not like common tigers, and must be sought for morning, noon, and night."—James Forsyth, *The Highlands of Central India*, 1871, 298

Logically enough, in India man-eating male tigers and leopards were often said to be manifestations or agents of Shiva in one of his many guises, and man-eating tigresses and leopardesses manifestations or agents of Kali in one of *her* various avatars. Durand recalled once hunting a man-eating tiger in central India. "The natives, of course, said that he was one of Shiva's horses," he wrote, "that he was therefore bullet-proof, and that no one could ever kill him." The sportsman failed to bring the killer to book.[1]

In the 1860s, while working as an assistant collector and magistrate in the Konkan — the coastal area of southwest Maharashtra between Mumbai and Goa — a young Englishman and typically avid sportsman called Arthur Crawford befriended a bespectacled old Brahman priest called Raghoba Mahadewrao in the town of Chiplun in the Ratnagiri district. Raghoba regularly visited Crawford's tent of an evening to read from his precious collection of tattered old Sanskrit manuscripts — documents that subsequently went missing when he became a sadhu, according to the Englishman — and one night he read out a tale dating from the middle of the seventeenth century: the story of the white tigers of the Koyna valley, which lies just east of Chiplun, between Kumbharli Ghat and the hill station of Mahabaleshwar, and has long since been flooded to form a reservoir.

According to this, when the mighty Maratha warrior Shivaji was a young man and lived with his recently widowed mother Jijabai at Partabgarh, near Mahabaleshwar — having been driven out of their palace at

Pune by the invading Moguls—two enormous tigers, a male and his mate, ravaged the valley for a great many years, killing people's precious cattle. (The idea that tigers form lasting pair bonds—widespread in colonial and former times—is now discredited by naturalists.) Their favorite lair was in the face of a cliff, under a huge rock on which they were so often seen basking in the sun—one always keeping watch while the other dozed—that it came to be known as the Tigers' Seat (Figure 7). Shikaris tried every

Figure 7. The Tigers' Seat at Mahabaleshwar, Maharashtra was the favorite lair of the legendary seventeenth-century man-eating white tigers of the Koyna valley, tigers said to be possessed by Shiva and Kali. It was later renamed Arthur's Seat, after Sir George Arthur, Governor of Bombay from 1842–1846, and is called that to this day (© Abhishek Joshi).

trick in the book to account for them. They dug cunningly concealed pits, but somehow the tigers always managed to scramble out. They smeared leaves with gum and lime and scattered them in the tigers' path, so that when the leaves stuck to their paws they licked them off and got the acrid mixture in their eyes. The male was almost blinded this way, but just as the shikaris were closing in the tigress led him to a pool to wash his eyes, and the pair made good their escape. Another time the roles were reversed when the tigress was caught in a drop-door trap baited with a calf: her companion set her free by gnawing through the wooden bars, then helped her carry off the calf.

One day, though, their luck ran out when they tried to separate a young buffalo from its herd and were both trampled on for their troubles, the tigress suffering broken ribs and her mate being gored in the withers. Together they just managed to crawl under a rock, where they collapsed. As the days passed the male's wound suppurated and filled with maggots, for it was out of reach of his tongue and the tigress was in such agony that she could not raise her head to lick it clean, and they wasted away until they were on the point of death. But then, somehow, they began to recover until one day they were strong enough to stand again. When they did so they found that their ordeal had taken its toll: their teeth were falling out, their eyesight was failing, their joints were stiff, and their muscles were weak. But, wonder of wonders, their coats had turned as white as snow, their once jet-black stripes showing but a pale orange in the sunlight. And they saw in each other's eyes a new craving: a hunger for human flesh.

From that day on they became the most terrible man-eaters in all the land. Men, women, children; all were fair game. Soon whole villages were abandoned, their cattle and crops left untended. No one knew where the white devils would strike next, for they soon grew strong enough to travel many miles between kills. All anyone knew was that almost every half moon they repaired to bask defiantly on the high rock and sleep off their latest meal.

Shivaji was away fighting at the time, and his mother did not want to bother him with so trivial a matter as the ravages of a couple of man-eaters. Instead she offered a huge reward of gold, land, and property to whoever accounted for the killers. But no one succeeded; perhaps no one even had the courage to try. Finally she was compelled to order two Maratha chiefs, Maloosray and Ghorepooray, to organize a hunting party and bring her the tigers' skins, or never show their faces in her court again.

The two chiefs led a small band of shikaris to a deserted village, where they set up base for the night in a hut. In deference to the man-eaters they

4. Fighting Back

allowed two Untouchables—Mhar tribesmen who had been left behind in the village—to share their accommodation, when normally they would have considered their presence defiling.

That very evening, as the hookah was being passed around the fire, the man-eaters burst in, seized the two Untouchables cowering in the corner, and carried them off into the night, the unfortunate pair screaming hideously as they dangled from the tigers' jaws.

Maloosray and Ghorepooray led the shikaris in hot pursuit, and they soon found the growling tigers settling down to eat their now mercifully dead victims by the side of a stream. Here the men hid until dawn, when the engorged animals slaked their thirst before curling up under a clump of bamboo to sleep off their meal. Signaling silently, Maloosray and Ghorepooray sent two shikaris back to the hut to fetch a rope net. On their return the shikaris sneaked up and threw this over the tigers, at which the two warriors, armed only with swords and shields, charged into the attack, yelling vengeance. The battle was brief but bloody. The male tiger mauled Maloosray, but soon fell under his ferocious assault. Ghorepooray, meanwhile, sliced open the tigress's belly, spilling her guts, but before he could finish her off she killed him with one swipe of a paw. She disposed of two shikaris in the same manner before crawling into the bamboo, where the remaining shikaris speared her to death.

After skinning the tigers and burning off their whiskers, Maloosray returned in triumph to Partabgarh, where Jijabai embraced him and hung her priceless necklace around his neck. News of the man-eaters' deaths spread fast, life in the valley returned to normal, and people flocked to give thanks at the temples of Bhawani, the Marathas' patron goddess.

Then Jijabai had a dream. Shiva appeared before her and said that he and Kali had been the real man-eaters. When the tigers lay dying under the rock he had sent them a peacock, a deer, and some monkeys to eat, and seven thrushes to pick the maggots out of the male's wound: hence their miraculous recovery. He and Kali had then possessed the tigers and begun their reign of terror. They did this, he explained, because for years Marathas had favored Bhawani over them. Everywhere their shrines lay neglected, not decked with garlands of flowers and anointed with fragrant oils, and everywhere their priests were in desperate poverty. Bhawani, it was true, had blessed Shivaji with a shining sword, but both Shivaji and she, Jijabai, appeared to have forgotten the very origin of the young warrior's name. Now, warned Shiva, a fearful famine would follow if this shameful state of affairs was not rectified immediately.

When she woke, Jijabai went straight to her astrologers, and on their advice held a great festival at Partabgarh in honor of Shiva, then another at nearby Satara in honor of Kali, feeding no fewer than 1,000 Brahmans at each. And from that day on until their final defeat by the British in 1818, Marathas honored Shiva, Kali, and Bhawani alike.[2]

White tigers have been recorded for centuries, and with their ghostly appearance have long been held in supernatural awe. In the Malay world, eminent individuals are often said to become benevolent white spirit-tigers post-mortem. According to Wessing, one such is Iskandar Muda, the great seventeenth-century sultan of Aceh, who is said to appear at his grave in Banda Aceh in the form of a white tiger every Thursday night, when people pray and make offerings to him there.[3]

In India there are many tales about white tigers. Burton wrote that the Gond tiger god was the Great White Tiger — a spectral tiger much larger than a real one — and just such a creature features in a tale told by Hanley in 1928; another story that features "purely fictitious" characters but was evidently based on real events, and one that would surely interest the most case-hardened psychiatrist.[4]

Following a bout of malaria, wrote Hanley, one of his bungalow servants on the Laojan Tea Estate, a Gond shikari called Lingo, went missing for five days until the planter found him collapsed in an old nursery garden at the jungle's edge. After several more days of delirium, Lingo recovered sufficiently to tell his tale.

On the day he disappeared, he said, he had felt very bad: so bad that he had drunk some liquor (for purely medicinal purposes). The next thing he recalled was waking in the jungle. And lo, there in front of him was a milk-white tiger with jet-black stripes, the most magnificent creature he had ever seen. Had he been carrying one of Hanley's guns he would have shot it on the spot. The tiger seemed to invite him to follow it: so he did, for mile after mile, until he found himself — under the brilliant light of a full moon — on the banks of the Pootinadi River in the burial grounds for coolie women who died in childbirth.

Here the tiger stopped, raised its head and roared, summoning forth the women's bhuts: particularly malevolent ones known as *churels*. A terrified Lingo heard the chuckle of their voices, but just then a cockerel crowed, and the spell was broken. Roaring in anger, the tiger moved on, and the Gond was compelled to follow, for a madness seemed to have taken hold of him. Days and nights passed until the weakening Gond found himself back on the banks of the Pootinadi, and the tiger once again unleashed an unearthly roar. This time there was no cockerel to save him, and in answer to the roar the churels of at least forty naked women with long flowing hair, small children on their hips, and flames shooting from their mouths—women many of whom Lingo had known—leapt out of the jungle and surrounded the Gond. "I was nearly mad with terror, Huzoor [Lord]," he told Hanley, "and then they set on me with sticks and whips ... and thrashed me unmercifully." At this point he must have passed out, for the next thing he remembered was waking up in the plantation hospital.[5]

Lingo clearly had a lucky escape, for according to Crooke, churels are notorious for seducing and abducting handsome young men, then keeping them until they grow old and all their friends have long since passed away.[6]

Black tigers have also been recorded, but are much rarer. Black leopards, meanwhile, are fairly common, especially in rainforests, where a dark coat is better camouflage than a light one, which is better suited to open, grassy areas. According to Wessing, black tigers are said to guard the graves of Lord Katong and Lord Meutoe, two warriors who helped found the sultanate of Daya, in Aceh, while Anderson recalled a livestock-lifting black leopard that he shot dead near Bangalore in the 1930s terrifying the locals, who assumed it was Satan come to harass them in animal form.[7]

Some man-eaters in India were said to be both black and white, depending on who you talked to, as Best discovered while stationed in Bilaspur. In 1908, a male tiger began eating Baigas and Banjaras in the Karidongri forest block in the area of the Chakmi Pass in the Lurmi Range there, sharing its victims with its "mate," and soon had the maximum 500-rupee government bounty on its head. One Baiga who Best knew, a noted archer called Chaitkoo, from the village of Boehra at the foot of the pass, took to carrying his bow and poisoned arrows everywhere he went, just in case; until the tiger pounced on him from behind. Best was planning a tour of the block in March 1909, and looked forward to tackling the man-eater, but with a few weeks to go he received a letter from an army officer—a gunner and veteran of the second Boer War (1899–1902) called Harry—requesting the exclusive shooting rights for the block for that month, a request that Best had no choice but to grant. He met up with the man at the old mud hut that served as the rest-house at the Gond village of Deosara, on the rough road winding through the forest-clad hills 3,000 feet up to the pass, took an instant liking to him, and after a couple of days shooting together proposed they join forces to tackle the man-eater. Harry readily agreed.

Their start was delayed for several days while Best dealt with frequent forest fires that he suspected the many bands of Banjaras who traveled up and down the road of starting deliberately to keep the man-eater at bay. But at last they got going, riding on elephants with a train of baggage camels and men, among them Best's trusty old Baiga shikari Amoli, armed with his usual ax and bow. "The Forest Ranger came too," wrote Best, "although climbing high passes on foot was not in his line; but there were fire lines to be inspected on top of the pass, and I had an unkind suspicion that the Ranger had never seen them."

These were 100-foot-wide fire-breaks cut through the forest by Baiga coolies. It was the ranger's duty to ensure they were kept clear by Baigas, who also sat in shifts on machans atop two-storey, thatched-roofed bamboo towers called *nakas* every few miles along each line, watching day and night for fires.

On their way up, Best and his party camped the night at Boehra where, reported the forest officer, the Baigas told him and Harry "the usual wild stories inevitably attached to all man-eaters. He was black; he was white! He had the head of a crocodile [as Shiva is sometimes portrayed] and the feet of a hyena. We did not feel compelled to believe any of these yarns."

The fire-watchers were particularly scared, so the next day when Best reached the head of the pass, where a fire line ran along the ridge, he gave all those in the naka there leave to return to Boehra.

That night he and Harry slept on the mud floor of the naka, while his men shared a dilapidated hut nearby, with the elephants and camels around the hut in a protective ring. On the final ascent Best had tethered buffalo calves to trees at intervals along the narrow path, using light rope, so that if the tiger killed one it could snap the rope and carry the carcass to a safe feeding place, giving him and Harry a drag to follow. (Another way was to use heavy rope, and a stout peg that only a tiger could yank out of the ground.) So now it was just a question of waiting. Night fell and all was still. Then, around ten o'clock, the tiger's roar shattered the calm, followed by the squeals of frightened men and animals. Circling around, the tiger roared again, as if in contempt, then moved off, growling, down the pass. The rest of the night passed uneventfully; though no one could say they enjoyed a sound sleep.

When Best and the others checked at first light they found the man-eater had killed one of the calves, dragged it under a bush, and half covered it with leaves. There were was no question of a beat: there were not enough

4. Fighting Back

men for one—"the tiger had settled that," noted the forest officer—and besides, the hill fell away in a cave-ridden cliff. So, after breakfast, Best and Harry tossed a coin to decide who would sit up for the tiger, and Harry won. Back at the kill, working in silence with the help of Amoli and a handful of Baigas, they constructed a machan by tying two poles—cut some distance away so as not to scare off the man-eater if it was still close by—across the fork of a sal tree, twenty feet above the ground, laying a charpoy on the poles, and placing a screen of green branches in front. Harry made himself as comfortable as he could on the charpoy, with his rifle across his knees, and a bottle of Scotch, but only one blanket, to keep him warm, the tiger being unlikely to return, if at all, until nightfall at the earliest.

Posting men to intercept any Banjaras coming up or down the path, Best spent the day checking the lines and catching up on paperwork. With a cold wind blowing, that evening he wore two sweaters over his pajamas as he enjoyed his after-dinner pipe, and was just thinking about bed and hoping Harry was still awake when two shots, followed by the booming alarm calls of monkeys, had him grabbing his rifle and yelling for his mahout. Within minutes he was heading down the path on elephant-back, the mahout carrying a lantern to light the way.

"What luck?" he called out.

"Go away!" yelled Harry.

"Why?"

"Because I have wounded the tiger and he is between us!"

Confirmation of this came when the elephant began tapping the ground with her trunk. Harry urged Best to leave, and the forest officer, who could only admire his friend's courage, reluctantly concurred. "The elephant's turn would have done credit to a polo pony," he reckoned. "She pretended not to be in a hurry up the path, as did the mahout. I am quite sure I failed to keep my dignity."

At sunrise Best found Harry none the worse for the night or the whiskey, and after a hasty breakfast they returned on foot with Amoli, covering him with their rifles. Amoli tracked a trail of blood to a bamboo thicket, where the man-eater lay dead of its wounds.

Two days later its hungry "mate" killed a Baiga in Boehra. This time it was Best's turn, and sitting over the remains he added her to the bag.

When the regulation six months had passed with no further killings reported, Best was able to claim the 500-rupee reward; and share it with an immaculately turned out and almost unrecognizable Harry when they met up again on leave in London in a gentlemen's club in Pall Mall.[8]

While traveling around Betul in 1862, Forsyth hunted a male tiger estimated to have claimed more than 100 victims in this largely Gond district. This brute and monster, as he described it, operated over such a large area — a wedge of country up to forty miles wide between the Moran and Ganjal rivers — that for a long time everyone thought there must be more than one killer on the prowl. When the young Scot arrived and pitched camp in a mango grove near the village of Lokartalae on the Moran River, the whole area was in the grip of fear. Many villages were deserted, while elsewhere everyone had barricaded themselves in their homes, only venturing outside when absolutely necessary. But even in armed groups, he learned, no one any longer dared travel on roads where the tiger had already claimed several lives. These included the main road from the teak forests of Betul to the railroad line being laid in the Narmada valley, with the result that the supply of ties had ceased.

As if that was not reason enough for Forsyth to bring the killer to account, he was besieged by people pleading with him to free them from its tyranny, a baby whose mother had been carried off while she drew water from a well even being brought into his tent and held up before him. Although they were too scared to hunt the tiger themselves, all the shikaris of the area gathered in his camp and offered their help. By this time the tiger had assumed wildly supernatural proportions in everyone's minds, the malevolent force behind it evidently being thought to be Shiva. Wrote Forsyth: "I was regaled with stories of the man-eater — of his fearful size and appearance, with belly pendent to the ground, and white moon on the top of his forehead [like Shiva]; his pork-butcher-like method of detaining a party of travellers while he rolled himself in the sand, and at last came up and inspected them all round, selecting the fattest; his power of transforming himself into an innocent-looking woodcutter, and calling or whistling through the woods till an unsuspecting victim approached; how the spirits of all his victims rode with him on his head, warning him of every danger and guiding him to the fatal ambush where a traveller would shortly pass."

Forsyth was hobbling with a sprained ankle at the time, but as soon as he could climb into a howdah he set off for the site of the latest kill, the village of Charkhera, on his favorite elephant, Sarju Parshad, a *mukna*, or tuskless male.[9]

Forsyth was as fond of Sarju as he was of Junglee, and rode him on many a tiger hunt. Clever and cheeky, Sarju could unfasten any rope or chain, and on dark nights, when no one was watching, would often roam around in search of food. He loved sugarcane best, and many times Forsyth had to dig deep into his own pocket to compensate farmers whose fields he had trampled. After such escapades he never allowed his keepers to catch him all the while they waved sticks and shouted threats, but surrendered the moment they resorted to persuasion and promised not to beat him. Once, when roaming free, he spied a lad sleeping with a sack of rice for a pillow. After pondering the problem for a while, he pulled the bag clear with his trunk while slipping the toes of one foot under the boy's head so as not to wake him. "Having gobbled up the rice with much despatch," wrote Forsyth, "he then rolled up the bag, and returning it under the boy's head stalked away!"

Now Forsyth was riding Sarju in the hunt for a man-eater. For once, he noted, his followers were tightly bunched around him, not straggling along behind. They comprised a guard of policemen brandishing muskets, a large band of matchlock-wielding shikaris, and some peons, or orderlies, carrying his spare guns, with baggage elephants in front and behind. Dense, scrubby teak closed in on either side of the path, below which wound a deep, narrow nullah — a dry river-bed, or ravine — studded with small pools and shaded by thick evergreen jamun — Indian blackberry — tamarisk, and karanda bushes: classic tiger-terrain in the dry season, when deciduous trees are bare.

On their way they passed several deserted villages, and numerous heaps of stones marking spots where the man-eater had claimed lives, passers-by having taken the precaution of adding a stone to each pile while offering up a prayer both to appease the gods and to protect themselves from the victim's bhut. (Variants of this age-old custom are recorded throughout Asia. In French Indo-China, noted Baudesson, passersby left a stone or twig in a box with a crude representation of a tiger carved in its side and placed in front of the main dwelling of a settlement ravaged by a man-eater, in honor of Duc-Thay.[10])

Forsyth's party found Charkhera deserted as well when they arrived and pitched camp there. That evening a messenger brought news that the man-eater had carried off a pilgrim on an isolated path near the village of Le on the Moran, so early the next morning Forsyth set out with two elephants and a few shikaris, accompanied by a young policeman detailed to inspect the remains. Where a small nullah running down to the main river crossed the path, the pole-suspended baskets in which the pilgrim had been carrying holy water lay in a pool of dried blood, and scraps of torn clothing showed where the tiger had dragged its victim through some bushes. Following the drag down into the nullah, they found his remains—a few bones and shreds of flesh, and his hands, feet and skull—in a thick stand of grass. Evidently the tiger had been disturbed at its meal. Its distinctive paw prints—the toe of one hindfoot dragged as a result of a wound inflicted by a shikari's matchlock some months earlier—led away through dense jungle. Trembling with fear, the shikaris tracked the killer on foot, while Forsyth followed on Sarju, covering them with his rifle. The trail ended in a jumble of boulders. The men lobbed in stones and firecrackers, but only a scrawny hyena emerged. With the sun soon to set it was time to head back. There was no point sitting over the remains: there were not enough left to tempt the tiger back, and anyway the typically cautious man-eater was known never to return to a kill.

Nearing the camp at dusk the party found the killer's fresh paw prints join the path. The men followed them warily in the gloom, and when "the Lalla" joked that the plump young policeman should be sent ahead as bait, he himself being far too thin to interest the tiger, he raised only a few hollow laughs. About a mile from camp the tracks turned off into a deep nullah.

That night Forsyth posted three elephants around the camp and ordered fires to be kept burning in between them, not that the men needed any urging. Early the next morning he rode out on Sarju and searched the nullah, but found no fresh tracks. He had just sat down to breakfast when some Banjaras came running into the camp to tell him the man-eater had just snatched one of their band near their camp at Dekna, only a mile and a half away. Sarju was still in harness, so Forsyth grabbed some food and an essential bottle of claret and was soon on his way with a few shikaris.

As they followed the drag through elephant grass almost as high as the howdah, Sarju stamped and trumpeted, and ahead of them the tall stems parted in a wave as the man-eater hurried off. Urged on by his mahout, Sarju followed as fast as he could, stepping nimbly over the half-eaten body of the Banjara. Only once did Forsyth actually see the tiger, and far too briefly for him to fire. Exiting the grass they tracked it for a mile along a nullah, through more high grass, then across some stony ground. They were heading further and further from camp, so Forsyth sent back for more shikaris and a small tent for himself. All day they trailed the man-eater, painstakingly tracking it paw print by paw print, but without catching another glimpse of it.

After spending the night in a village by the Ganjal, at sunrise they tracked the tiger along the dry, sandy river-bed and into a dense clump

4. Fighting Back

of jamun and tamarisk near the village of Bhadugaon. No tracks led out again. With no other water or shade in the area, Forsyth reasoned the tiger would stay inside this cover all day, so he led his men to Bhadugaon for a welcome breakfast.

At about eleven that morning they silently set out again. With men posted in trees all around, and an elephant blocking the only escape route up onto the high river bank, Sarju crept into the tangle of roots and dangling branches, the Lalla sitting behind Forsyth in the howdah. On reaching the middle the elephant stopped, and began kicking up the earth and growling in a low rumble: the man-eater was somewhere close by. "We peered all about with nervous beatings of the heart," recalled the young Scot. The mahout whispered that he could see the tiger lying under a bush. At Forsyth's command, the Lalla threw in a stone; and out shot the man-eater, bounding away with a roar towards the river bank. Forsyth gave it both barrels, but though wounded it kept on going; until it saw the other elephant, at which it turned and charged straight at them with another roar. It was within twenty yards when Forsyth fired again, bowling it over, but it picked itself up and charged again. Forsyth was just about to fire for a third time when Sarju spun round and he found himself facing the wrong way. The next moment the roaring tiger leapt onto Sarju's rump with outstretched claws, and the mahout's terrified boy fell clear, leaving Forsyth, the Lalla, and the mahout struggling to cling on as Sarju kicked out at his attacker. Finally Sarju paused long enough for Forsyth to lean over, put the muzzle of his large-shelled rifle to the tiger's head and blow its skull to smithereens (Figure 8). As he recalled: "He dropped like a sack of potatoes; and then I saw the dastardly mahout urging the elephant to run out of the cover. An application of my gun-stock to his head, however, reversed the engine; and Sarju, coming round with the utmost willingness, trumpeted a shrill note of defiance, and rushing upon his prostrate foe commenced a war-dance on his body."

Sarju had been badly mauled, but only two days later Forsyth took him out again and bagged two more tigers without him so much as flinching.

As for the man-eater, "He had no moon on his head," wrote Forsyth, "nor did his belly nearly touch the ground. I afterwards found that these characteristics are attributed to all man-eaters by the credulous people."[11]

In Mandla in 1889, in a part of the district known as the "Chaurassi Zamindari," or the "Zamindari of 84 villages," a zamindari being an area controlled by local landlords, Hicks hunted a man-eater that killed 119

Figure 8. An artist's impression of pioneering forester James Forsyth, on his favorite elephant, Sarju, putting paid to a man-eating tiger, said to be an agent or manifestation of Shiva, in the dry bed of the Moran River in Betul in 1862. Top left is the elephant positioned to prevent the tiger escaping up on to the high river bank, top right one of the men posted in trees as stops. Center left the mahout's boy flees for his life having fallen off Sarju. With Forsyth in the howdah is his faithful shikari, "the Lalla," while in front is the mahout, a "miserable opium-eating villain" (Frontispiece to Forsyth's *The Highlands of Central India*. London: Chapman & Hall, 1871. Courtesy of Bloomsbury Auctions).

people—most of them Gonds—in the space of a year: a demon, as he himself called it, that everyone for miles around said was Bhawani in the form of a tigress. It spread such fear that village after village was abandoned to be overgrown with jungle. Government forestry work came to a halt, all the local coolies having fled, and Hicks had to send in outside labor; but before long these men also bolted.

Despite this, Hicks had no official reason for visiting the area, but he was intent on ridding it of the pest, so as soon as the rainy season ended he set off with his wife and two young daughters, using a government and zamindari forest boundary dispute as an excuse. On arriving and setting up camp he was struck by the height of the villagers' field machans—platforms from which they guarded their crops against raiding deer and wild boar at night—because at some twenty-four feet off the ground they were twice as high as normal. He was also struck by the spacing of their rungs, which were a good six feet apart. Told that the man-eater frequently car-

4. Fighting Back

ried off watchers from beneath the thatched roofs of even these towering structures, he felt sure that such an agile killer could only be a leopard.

This was confirmed when he was shown the remains of a victim killed ten days earlier: a few bloodstains, and scattered scraps of clothing, hair, and bone on the ground; and, wedged in the fork of a tree fifteen feet above the ground, part of the skeleton, picked clean by crows.

A few days later, when Hicks was passing through a village, a man came up to him and said that about a week earlier "some animal," as he put it, had snatched his wife from their hut in the middle of the night; which prompted Hicks's mischievous old shikari to ask if it had been a two-legged animal. Further inquiry revealed that the man-eater had forced its way through the thatched roof, killed the woman in front of her cowering husband, then torn an exit hole in the bamboo-matting back door before dragging her off into the night.

On his way back to camp, a machan on which the man-eater had mauled a watcher's leg was pointed out to Hicks, who went to see the man and hear his story. Lying on his bed, feverish from his infected wounds, he told Hicks he had spent the night on the platform with six companions. None was armed, but as usual they kept a small fire burning in an earthenware pot, from which they lit their hookahs, until tiredness overcame them and they lay down to sleep. Cramped for room, the man let one leg dangle over the side. During the night the man-eater seized the limb and tried to pull him from his perch, but he clung on, and on being woken by his screams one of his companions was quick-witted enough to empty the hot ashes from the pot over the leopard's head, at which the leopard let go and fled. But a few hours later it came back, and the men spent the rest of the night fending it off. Hicks did what he could for the man, who refused to go to hospital, by bandaging up his leg, which was terribly swollen, but the forest officer knew that as soon as he left, the man would remove the bandages and fill the wounds with the usual mixture of slaked lime, tobacco, and cow dung. He was not surprised when he later heard that the man had died soon after his visit.

Three Gond shikaris then told Hicks how one night the leopard had killed a companion as they sat up a tree over a man's remains. They swore blind that as the man-eater approached its meal the corpse raised an arm and pointed straight at them. Hicks reckoned that most likely the gruesomeness of their task, with this familiar superstition foremost in their minds, got so much on their nerves that they imagined seeing what they half-expected to see. Whatever the case, the four men, all armed, froze in terror when the leopard arrived, allowing it to climb their tree and drag one of them away at its leisure.

Meanwhile, every few days reports reached Hicks of a fresh killing.

Near Dadargaon and Chikly the leopard snatched a man as he slept in his makeshift shelter in a clearing he had planted with wheat. A flattened patch of wheat showed where the leopard lay swishing its tail while it waited for him to fall asleep before pouncing. Paw prints told Hicks that the killer was a leopardess, and though he knew it was pointless he beat the area to please the villagers.

A 16-year-old girl spent her nights alone in a bamboo tepee-like shelter, guarding her family's crops by tugging on a string tied to rattles outside in the field. One night the rattling stopped: the leopardess had claimed another victim.

An old man and his son traveling through the area slept side by side one night under a village's sacred peepul tree. The man-eater carried off the son without waking the father. Grief-stricken, the man took his own life.

But each new kill was always several miles from the last, and news of it always took a day to reach Hicks, by which time the killer could have been anywhere. So rather than chase after it blind, he waited. "It was [a] very trying time for us all in camp," he recalled, "for none knew at what moment this lurking fiend might pounce upon and kill one of our number, especially at night, camped as we often were with dense jungle growth almost up to our tents."

Each night he ordered fires to be kept burning around the camp, and armed orderlies to patrol its perimeter. One night he found one guard asleep, so to teach him a lesson he sprang on him with a roar and dragged him by his feet into the jungle. But he had not reckoned on what would happen next. The man started yelling *"sher! sher!"*—"tiger! tiger!"—and instantly the whole camp was in uproar. Buckshot and bullets whistled past their ears, and Hicks had to pin the guard down to prevent him from being hit. When the Englishman finally restored order, the man thanked the sahib for saving him from the man-eater.

Another time, Hicks and his wife had a fright of their own. They slept in adjoining beds with their baby girl between them, and Hicks had taken to putting his bed across the entrance to the tent, tying six of his best dogs to it, and sleeping with a loaded rifle and spare cartridges within easy reach. As he recalled: "One night my wife sat up with a gasp and said, 'Baby is gone!' What an awful sensation it was: for a moment, frozen with

4. Fighting Back

horror, we stared at each other with despairing faces, and the next moment were frantically searching the bedding, turning over the mattresses, but all to no purpose — the child was gone! Was the man-eater then really a supernatural ghoul, which the natives said it was? How then could it have entered our tent without our knowing it?" Tearing the beds apart, Hicks found the infant asleep on the floor, her thumb still in her mouth: she had slipped down the gap in the middle.

One morning came news that the killer was trapped in a cowshed. Hicks found a crowd gathered around, the men all urging one another to go inside and kill the shaitan before it escaped. But it already had, through a hole in the roof.

Its days were now numbered, however. Hicks found its fresh paw prints on a track where there was also a familiar stench, one he had last smelt in Mysore during the terrible famine of 1876, when the pye-dogs lived on nothing but corpses: the stench of dung from an animal that lived on human flesh. Sure enough, he found steaming piles buried under leaves: he was hot on the trail. Moreover, there were two sets of paw prints, telling him the killer shared her spoils with a "mate." (Hicks doubted that the male did any killing, reckoning "females are always the more daring and vindictive in such matters, and [she] would probably not trust him to carry out such important business while she was by.")

The trail led to the outskirts of a village where the leopardess had killed a man in front of his young wife two weeks earlier. Each evening since, said the headman, all the villagers had gathered at his hut, where they beat drums all night long to keep the shaitan at bay, for none dared sleep in their own home. The widow was brought forward, and seemed rather to enjoy being the center of attention as she told Hicks all about how her husband had been killed.

The villagers were keen to beat for the killer, but Hicks knew this would only scare her away, so instead he ordered a sty to be built just outside the village, with three walls of stout posts sunk deep into the ground, and a roof of equally strong posts, so that if the killer was suspicious of the open doorway she would not be able to dig her way in from the side or break through the roof. That done he had thorns piled all over and around the structure until it looked like a bush. At sunset he tethered a piglet inside, then across the entrance he stretched a series of strings attached to a bar from which a single string wound around a peg to the trigger of a loaded rifle mounted on forked sticks and pointed at the doorway. Much to his satisfaction, he recalled, the piglet played its part, its frantic squeals echoing through the still night air. As they left the scene Hicks and his men talked loudly all the way so that if the leopards were watching they would know the piglet was now all alone. Ordering silence in the village for the rest of the night, Hicks retired to camp to join his wife for dinner.

"We were just raising our first spoonful of soup to our mouth," he recalled, "when bang went the gun." Grabbing a rifle he ran ahead of a crowd of excited villagers and orderlies brandishing spears and flaming torches. At first they could see nothing. They pelted the sty with stones, at which the piglet squealed furiously and everyone momentarily retreated in panic. Then, barely visible in the torchlight, they made out the form of a leopard in front of the trap. The killer was dead, shot clean through the head. Examining her, Hicks found she was an old animal with broken teeth.

"The villagers were simply intoxicated with delight," he wrote, "and danced and capered round me and my wife, calling down blessings on her head, a number of them prostrating themselves and embracing her feet, thanking her as their goddess who had come and saved them from the scourge of the terrible 'Bhawani' who had devastated them for so long."

Hicks reckoned that after a few months the craving of the dead killer's "mate" for human flesh would make it turn to killing its own human prey, and sure enough a few months later it killed an old woman collecting firewood. When a cowherd met the same fate two weeks later, four Gond shikaris screwed up the courage to sit up in a tree over the man's remains, each armed with a matchlock loaded with bits of scrap metal. When the leopard came they fired as one, then huddled together in terror for the rest of the night. In the morning they found the animal dead on the ground below. Three of them had missed it completely, but a piece of old telegraph wire from one man's gun had gone straight through one ear and into its brain.[12]

5
Shapeshifters All: Like Weretiger, Like Werewolf

> *"Tales of this kind should be told, as in India, in the evening shadows under the village peepul tree, suggestively whispering of ghost-land overhead, while the vast background of the outer dark beckons the fancy to a far travel. Under these circumstances the absurdity of animal transformation assumes a dignity and reasonableness impossible to convey in print."* — John Lockwood Kipling, *Beast and Man in India*, 1891, 358

Practitioners of the *harimau*, or tiger, style of the ancient Malay martial art of silat (Figure 9) mimic the movements of tigers. According to Wessing, some silat masters invite the tiger-form spirits of dead teachers to possess and empower them, some people in Aceh even saying there are silat masters who turn into tigers mid-fight as a result.[1]

Physical shapeshifting is one kind of werebeastery. Another is spiritual shapeshifting: when you send out your shadow soul to enter and control a real animal. When a Batek shaman trances or sleeps, heard Endicott, he sends his shadow soul deep into the forest to wake and enter a special tiger body given to him by an ancestor, then prowls around, visiting Batu Balok, patrolling his own camp, and eavesdropping on others.[2]

Physical or spiritual, not all shapeshifting is voluntary. Uninvited spirit possession of one kind or another can result in involuntary physical transformation. De Groot wrote of a palace dignitary called Shi Tao-süen in Hubei province in the fourth century who one day foolishly told a gathering how he became so deranged in his youth that for a whole year he regularly changed into a tiger, in which form he killed and ate countless victims, stealing their valuables for good measure. Some of his victims' relatives were listening, and they immediately seized him and handed him

Figure 9. The Malay martial art of silat originated in Sumatra. This photograph of a demonstration by two Minangkabau silat masters — the Minangkabau being the largest ethnic group in Sumatra, originally from the western highlands there — was taken some time between 1900 and 1930 (Tropenmuseum, Amsterdam [60023125]).

over to the magistrates. They in turn had him thrown in jail, where he starved to death.[3]

In medical terms, the delusion you are a wolf, tiger, or other such fierce animal, and accordingly behave as such, is called lycanthropy. Witnesses may themselves be convinced. In Kemaman, a Malay told Locke he was once in the jungle with two friends when they came across a wild-looking man on his hands and knees and frothing at the mouth. He growled so fiercely at them, they ran away. "Clearly," the Malay said, "the man was in the act of transforming himself into a tiger. We were fortunate to escape from the spot unharmed." In Pante Cermen in Aceh, Wessing heard of a man in the early 1970s who, as well as shaking and jumping about a lot, apparently sprouted fur and claws when enraged. It first happened when a tiger ate his brother and he got angry thinking about it, then happened several times subsequently when fellow villagers upset him in some way; until they learned not to. Each time they had to roll him up in a rattan mat to calm him down and return him to normal.[4]

Rage presumably also triggered the supposed transformation — a permanent one, this time — in the case of a man-eater Burton hunted in Berar in 1891: a leopard said to be a man who transformed when his wife threw a stone at him.[5]

To gain audience sympathy, the cinematic werewolf is often the victim of an affliction inflicted by a bite — saliva being a well-known repository of "soul stuff" — or by a hereditary curse. Similarly, in parts of the Malay world, reported Wessing, it is said there are people who, as well as becoming tigers of some sort post-mortem, are involuntary physical weretigers as part of the price agreed by an ancestor for acquiring wealth and power through "tiger magic." A case in point are the people of Daya who, as well as being said to transform post-mortem if washed with diluted lime juice according to Islamic custom, are said to transform while alive if they accidentally smell limes or lime juice, not resuming their human form until they have had their fill of raw beef or human flesh.[6]

Meanwhile, numerous Sema Nagas told Hutton they were, or once were, involuntary spiritual shapeshifters: usually wereleopards, but occasionally weretigers. No one ever wanted to be one, for the habit was dangerous, exhausting, and greatly to be feared. Luckily, though, it could be shaken off in old age.

Sema Naga werebeasts were mainly men, and mainly livestock-lifters, though men who had taken human heads killed as many men in their animal form, and the few female wereleopards were even more dangerous. The affliction struck at the bidding of spirits, usually when you had been alone in the jungle for a few days. It could also be acquired by intimate association with someone already afflicted: by never leaving their side for months at a time, sleeping in their bed, or — cue the role of saliva — sharing their food or drink. When the person slept, then occasionally their shadow soul went out of its own accord into their animal body — their *mavi* — usually for the night, but sometimes for days. The sleeper — or dreamer, if you prefer — twitched convulsively, mirroring the mavi's movements: one man once bit off one of his wife's breasts as his leopard made a kill. By day, if their shadow soul was still abroad, they went about their business, but in a lethargic, mumbling, zombie-like way. If their leopard was being hunted they felt its fear, and again mirrored its movements, leaping about wildly as it tried to escape.[7]

When people need help, reported Carey, a Kensiu shaman holds a *panok* ceremony, entering a palm-leaf hut in a clearing at night while everyone gathers round. Whistles and slaps come from inside, then the rustle of leaves: helpful spirits have arrived. As they possess the shaman in turn, he speaks, chants, and sings in their different voices. Finally, he roars: *bidog*, the tiger, has possessed him, ready to answer people's questions with oracles.[8]

According to Carey, the Kensiu genuinely believe the shaman turns into a tiger. The anthropologist stressed he implied no deception, the shaman believing it too. But as Wessing points out, shamans practice their white magic in a bounded, controlled, and civilized environment — like a purpose-built hut, indeed — and so the forces they invoke are constrained. The Kensui shaman behaves like a tiger, but that is all. Likewise a silat master is imbued with the tiger's power, nothing more. But when a black magician — a witch or sorcerer — invokes the same forces in an unbounded, uncontrolled, and uncivilized environment, such as deep in the forest at night, the forces are unconstrained: and the result, say some, is actual physical transformation.[9]

Spiritual or physical, involuntary werebeasts are barely, if at all, conscious of their actions in their animal form — they may kill you and eat you, but it is nothing personal, you understand — whereas voluntary ones remain fully self-aware. You might think, then, that a black magician would need a pretty good reason to undergo such a momentous experience as physically becoming an animal. After all, the very idea represents crossing a fundamental boundary. Yet sometimes the purported motive is merely pecuniary gain. In 1986, it was reported that a mob of Malaysians beat a dog to death because, they said, it was a member of a gang known to commit robberies in animal form.[10]

Becoming an animal just to commit robbery is rather drastic, to say the least. Much more fitting is to do it to unleash your repressed atavistic desires. Edgar Thurston reported that in the Malabar region of southern India it was said — no doubt scurrilously — that a Paniyan tribesman lusting after a particular woman goes to her home at night with a hollow bamboo, walks around the dwelling three times to draw her out, then turns himself into a bull or dog, in which form he "works his wicked will." (Within a few days, unsurprisingly, she dies.)[11]

Vengeance is a common motive, as we shall see, and often, vengeful shapeshifters do not just kill their enemies, they devour them body and soul. This leads on to another classic motive for willing werebeastery, spiritual or physical: a craving for human flesh and blood. According to Wessing, black magician *tjindakus* or *tjindaks*—weretigers—in the Tapanuli Selatan area of North Sumatra knock on your door in the evening and ask for a bed for the night. If you are foolish enough to fall for this classic ruse

5. Shapeshifters All

they wait until you have fallen asleep then turn into tigers and suck your blood. And if you are really unlucky they devour your heart, lungs, and liver too.[12]

In Aceh, heard the anthropologist, such weretigers are more discerning, devouring only the liver or heart. Like Daya's involuntary weretigers, they must not smell limes or lime juice, or they too involuntarily transform. Typically they live in the mountain areas of the province. Anyone going to Tangse, for one, must be careful not to go with evil intentions towards its weretigers, because they will somehow always learn of your plans. When five men once went into the forest there and one said that if they met a weretiger he would kill it, a tiger or weretiger — no one was quite sure which — sprang out of nowhere and killed him on the spot. And when a weretiger once killed another such man in Tangse, his friends buried him in a four-meter-deep grave and piled stones on top, but that night the weretiger still came and dug him up.[13]

Black magicians sometimes turn other people into animals for their own nefarious ends. Crooke noted that outsiders said all Tharu women were witches who turned strangers into wild animals, while Elwin reported Gonds often saying that Baigas could turn someone into a fox or jackal.[14]

Presumably, though, black magicians in Asia sometimes turn other people into prey animals, then themselves into predators to kill and eat them. For certain, in Ivory Coast in West Africa, reported Swedish anthropologist Birger Lindskog in 1954, black magicians were said to turn their victims into prey animals such as antelopes, then themselves into leopards or hyenas to hunt them down.[15]

And ideally for the magicians, you would think, the victims, unlike other involuntary shapeshifters, must remain horribly self-aware in their animal form, in the tradition of the huntsman Actaeon in Greek mythology. Actaeon came to grief when he offended Artemis, the virgin goddess of fertility and hunting who had a wolf emblazoned on her shield and was often identified with two primitive, pre–Greek deities: Selene, a goddess of the moon; and Hecate, a goddess of childbirth, crossroads, the underworld, and the moon, and the patron goddess of Western witchcraft. Artemis turned Actaeon into a stag for seeing her bathing with her nymphs. Chased by his followers and hounds, he tried to shout out to them, but no words came from his animal mouth, and the hounds tore him to pieces, urged on by his men.

Not all Malay weretigers who physically transform at will are bad news: not all of the time, at least. In 1894, Jacobs told of a rattan-cutter in Aceh who went too deep into the forest and became lost. Eventually he spied a hut and knocked on the door. A man welcomed him inside, and the rattan-cutter asked if he could spare any food. Telling him to wait, the man said he would fetch some fresh meat. Just then the rattan-cutter noticed the man's feet were crooked: a classic weretiger trait, just as joined eyebrows are the mark of a werewolf. Sure enough, once outside, with some shaking movements — much like the man Wessing heard about in Pante Cermen, indeed, though in that case the transformation was involuntary — the man turned into a tiger then bounded away. Shortly after, he returned with a water buffalo, turned back into a man, and presented the meat to his terrified guest.[16]

Crooked feet, or the absence of heels, also identified weretigers — or tigers turned into men — in Hubei, reported de Groot. (Indeed, they are a common feature of many mythical creatures, not least among them, according to Benedict Allen, the so-called Gugu or apemen of Sumatra, as well as, wrote Crooke, churels.) In 1899, Dutch scholars Josef Knebel and Godard Hazeu reported that other features identifying a weretiger — physical or spiritual, voluntary or involuntary — in the Malay world, appearing after the first shapeshift, and remaining thereafter, are the lack of a philtrum (the cleft below the nose), giving a smooth upper lip, and gaze aversion, or a shamefaced inability to look you in the eye.[17]

At least, that is what people say. Ton Schilling, a young Public Works Department employee in the Palembang district of South Sumatra before World War II, told of an old widow who lived in a hut in a clearing in the Anai Ravine, a remote area of then virgin forest in West Sumatra also known as the Tiger Ravine, so many tigers did it hold: a hut not built on stilts in the typical Malay manner, despite the number of tigers around. She told a Belgian army officer she befriended, a man called Loubelle, that she had been thrown out of her village for marrying a tigerman, but had been happy, and that her three grown-up sons were all tigermen too and lived in the forest: she seldom saw them but they were good boys who ensured she never went hungry. When Loubelle accepted her offer to "fetch" him a stag to shoot, she hollered across the ravine, and a succession of stags and wild boar hurtled out of the jungle, driven, she explained, by her sons: "They had turned into tigers, and the rest was easy." Schilling himself befriended her and enjoyed a successful hunt led by one of her boys: a young man with a frank, searching gaze, and a normal philtrum.[18]

Worldwide, a common means of inducing physical transformation is swallowing some magic substance or other, in which case, as we shall see in stories from India, an antidote is often needed to reverse the transformation; one you may have to entrust to an accomplice.

In China, a tiger skin does the trick. De Groot told of a ch'ang kwei that transformed a lone mountain-traveler by throwing such a skin over his head. Under the domination of the spirit, the poor fellow became a terrible man-eater for several years, until one day he managed to escape its clutches by running into a Buddhist monastery and hiding under a bed. One of the monks asked of him, "My dear disciple, what do you want here? Is it to devour us, or are you merely disguised as a beast?" At this, the weretiger lowered its ears and wept. Taking pity on him, the monk cared for him from that day forth, and within six months he had resumed his human form. After two years of not daring to step foot outside the monastery he began venturing a short distance into the outside world again, until one day he found the spirit waiting for him. Once again, the spirit threw a tiger skin at his head, but this time its aim was off, and the skin only brushed his legs, which again assumed tiger form. Back in the sanctuary of the monastery, he vowed to recite holy books for over a year, after which he regained his full human form. And he never set foot outside the monastery again.[19]

One way to become a tiger in Java, reported Skeat, is to don not an actual tiger skin, but merely a tiny, magically elastic tiger-striped sarong, one that initially covers only the toes but then somehow stretches to engulf the whole body; or, if you are in Sunda, heard Wessing — though it only works outside at night — a similarly titchy and magically elastic shirt cut from a weretiger's funerary shroud.[20]

Another favored method in Java, according to a 1902 report by Dutch translator Johannes Winter — one actually written in 1824, noted Boomgaard — was to pray then jump and roll about. Elaborating on this, in 1899 both Knebel and Hazeu described the actual method as to recite certain formulae, or turn your shirt inside out and hold your breath, then — a bit like the shapeshifting wolf in some versions of the Russian fairy tale about the Firebird — tumble head over heels three times. Such a "somersault tiger" sleeps deeply by day, wrote Banner, then sneaks out to a secret spot in the jungle. Midway through his somersault — Banner specifies only one — a tail and stripes sprout forth as his forehead touches the ground, and on landing he is transformed. After spending the night prowling and satisfying his hunger, just before dawn he returns to the spot and reverses the transformation with a single backwards somersault.[21]

Wessing linked the somersault to a spiral or swastika, representing a whirlpool and therefore a symbol in many cultures of the watery interface between this world and the spirit or underworld. Indeed, water features in shapeshifting lore worldwide as a medium of transition, not least in European lore. Artemis turned Actaeon into a stag by splashing water on him. In the Malay world, as we have seen, tigers become men and men tigers by swimming across or passing under a river or lake. Similarly, first-century A.D. Roman writer Pliny the Elder told how young men of the family of Antaeus drew lots, the loser to become a wolf by swimming across a lake. Nine years later, provided he had not tasted human flesh, he could reverse the transformation by swimming back again. Read almost any werewolf author — Hurwood, for example — and you will come across reference to people saying drinking from the same stream as a wolf, or from rainwater puddled in a wolf's paw print, turns you into a wolf. Similarly, Angami Nagas told Hutton there was a spring — some even said a spring of blood — somewhere in Sema Naga country that turned you into an involuntary spiritual wereleopard or weretiger if you drank from it. People sometimes made crying babies drink from the spring: they stopped crying, but grew up to be such werebeasts.[22]

Pliny's young man first took off his clothes and hung them on a tree. Adam Douglas argues that clothes are such a key part of what distinguishes humans from animals that no voluntary transformation from human to animal is possible until they have been removed, and no reversal of the transformation can occur until they have been put on again. This requires their protection by magic, an idea dating to at least the time of Pliny, whose contemporary Petronius imagined a slave called Niceros see a soldier friend become a wolf one night after stripping off then urinating around his clothes, so turning them to stone until his return.[23]

Winter and Knebel both specified that a "somersault tiger" similarly cannot transform without first removing his clothes, which he then hides, as he cannot resume human form without them. Even Daya's involuntary weretigers take off their clothes, unawares, when they begin to transform, noted Wessing.[24]

I am not so sure. Disrobing, then hiding the clothes — or even turning them to stone — may simply be practical measures. Anyway, in some stories the clothing clearly changes with the body; the magically stretchy shirt or sarong goes on over your normal clothes, for example.

Hazeu wrote that spitting sirih, which contains limes or lime juice, on a weretiger in tiger form in Java makes it resume its human shape. And to snap a West Javan silat master out of his trance, heard Wessing, you must both spit on him and ritually sprinkle him with water.[25]

But an even simpler method of making a physical werebeast of any sort "resume their human form"—that is, snap out of their trance—or otherwise render them harmless is to address them directly by name. This is a worldwide phenomenon, one based on the wider idea that a person's evil spirit—be it their ghost or their wandering shadow soul—cannot harm you if you know their name. Wessing heard it said of the weretigers of Daya, while de Groot reported it common in China, citing tell of a Han dynasty (206 B.C.–A.D. 220) prefect in Anhui province, a man called Fung Shao, who "one day turned into a tiger, and devoured the people of his jurisdiction," until a crowd shouted at his tiger form, "Lord-Envoy Fung!" at which he ran off and was seen no more. The sinologist also told of a man called Ch'en Shih-san who nightly turned himself into a livestock-lifting tiger by donning a tiger skin. Benefiting from the fresh meat, his relatives turned a blind eye until one day his sister-in-law caught him returning home with a human leg slung over his shoulder. He dived under his tiger skin, turned back into a tiger, and from that day on lived in the mountains as a man-eater. But if those who knew him met him, all they had to say was, "Ch'en Shih-san, old man, I am your neighbour, harm me not," and he would droop his ears and tail, and slink away.[26]

In Robert Masters' and Jean Houston's celebrated 1960s work *The Varieties of Psychedelic Experience*, "S-6," a 37-year-old anthropologist, told how he chemically induced the hallucination of turning into a tiger. Putting on some appropriately ritualistic music, he took some LSD and lay down to await developments. Only minutes later he found himself crawling around his apartment, spitting and snarling and believing he was a tiger. Looking in a mirror he actually saw a tiger; albeit one with an oddly familiar face. When the effects of the drug wore off he regretted having to leave his tiger self behind, even though he had not been a happy tiger. Asking himself if he would have looked in any way like a tiger to anyone watching, part of him thought yes.[27]

When people say they have seen someone turn into an animal, then either they are lying, are mistaken, were deceived, dreamt it, or were themselves hallucinating. But when you believe in witchcraft, the power of suggestion alone can make it work. Numerous anecdotes tell of people who, convinced someone has put a fatal spell on them, resign themselves to dying when medically nothing is wrong with them or because—like the Muong who believed he was the victim of tiger-whisker poisoning—they refuse help. Likewise an observer who believes a transformation can happen may actually see it happen. As Maria Uribe, director of the Colombian

National Institute of Anthropology, put it in 1999: "Maybe seeing is not believing but vice versa. You can't see unless you believe. Accept that, and reality has few bounds."[28]

When attacked, learned Bakels, Kerinci dukuns may call on their spirit-tiger familiars, who are silat masters, for help. The spirit tiger then either appears and confronts the attacker itself, or it possesses the shaman, in which case the attacker sees him "as if he were a tiger."[29]

This, then, might explain why some people say some silat masters actually become tigers mid-fight.

In the Malay Peninsula, noted Wessing, a weretiger is a *rimau jadi-jadian*, *rimau* meaning "tiger," and *jadi-jadian* "imitation," while in Java a weretiger is a *macan gadongan* (or *gadungan*), *macan* meaning "tiger," and *gadongan* or *gadungan* "false" or "disguised."[30]

This hints, to say the very least, at the real nature of the phenomenon.

Transformations classically take place at night, when predators and evil spirits are on the prowl and all good people are asleep in their beds. The trouble is, night-time is dark and often foggy, which may be atmospheric on screen, but is inconveniently obscuring in real life. The solution is moonlit metamorphosis; and ideally metamorphosis under the brilliant light of a full moon. This is a staple motif in shapeshifting lore dating to at least Petronius, who had Niceros see his friend transform when "the moon was shining like one o'clock."[31]

In Europe the moon has fundamental ancient associations with magic, hunting, fertility, and madness, all of which can be linked as a possible original source of werewolf lore. In outline, the idea is that for prehistoric hunter-gatherers, for whom fresh meat was precious protein, the full moon was the signal for the men to don the skins of their totem animal the wolf and work themselves up into the literally lunatic frenzy required for the hunt. As Douglas points out, this especially applied to the hunter's moon, the first full moon after the fall harvest moon, winter being when fresh food is scarcest. British anthropologist Chris Knight has even proposed that women cunningly coordinated their menstrual cycles so that their periods coincided with the occasion, thereby depriving the men of their conjugal rights until they had damn well filled the larder.[32]

In werewolf fiction and films, a full moon often triggers involuntary transformation, but contrary to popular belief there is no actual link between abnormal behavior and phases of the moon.[33]

In India, the power to shapeshift is bestowed by such powerful deities

5. Shapeshifters All

as Shiva, who is typically depicted with a crescent moon on his head, the bloodthirsty Kali, in the form of Durga or Bhawani, and the mighty Earth Mother, who is often identified with Kali and is very much the equivalent of Artemis. But really lunar illumination is primarily a narrative device for making a nocturnal transformation visible — just as it helps to have a witch ride her broomstick across the face of a full moon — and I suspect that in colonial weretiger stories full moons are simply put in for effect by authors more familiar with European werewolf yarns. Sometimes, indeed — as Hutton noted in the case of Sema Naga wereleopards — shapeshifting is actually said to be commonest between the expiry of the old moon and the rising of the new.[34]

This may be because attacks by real leopards are more likely to succeed then. For sure, a 2011 report by Craig Packer, Alexandra Swanson, and Hadas Kushnir, all of the University of Minnesota, and Dennis Ikanda of the Tanzanian Wildlife Research Institute, shows that most lion attacks in Tanzania occur when the moon is waning: when lions can surprise people and other prey following a period of too bright moonlight. Its authors quite reasonably say that this could explain the full moon being a harbinger of doom.[35]

Another staple motif of shapeshifting lore is repercussion wounding, or wound doubling: that is, any wounds inflicted on the animal form are duplicated on the human, and vice versa. De Groot told how a man in China stayed out late one evening collecting firewood, and when it grew dark two tigers chased him up a tree. Unable to reach him, the tigers said to each other, "If we can find Chu Tu-shi, we are sure to get this man." One then went away, and soon after a longer, leaner tiger appeared and made a grab for the man, but luckily the moon was shining so he saw the paw and hacked at it with his ax, at which both tigers fled. On returning home the next morning he relayed his adventure to his fellow villagers. A posse descended on Chu Tu-shi's house but were told they could not see him as he had hurt his hand. So the gang went to the district prefect, who sent his underlings, all armed with swords, to set fire to the property. At this, Chu Tu-shi leapt out of bed, turned into a tiger, and escaped through the crowd, never to be seen again.[36]

Again, in European werewolf lore wound doubling dates to at least Petronius, who had Niceros claim that the morning after he saw his friend become a wolf he learned that someone had put a spear through a wolf's neck in the night, then found his friend prostrate in bed with a doctor seeing to his neck.[37]

The risk of wound doubling is not restricted to physical shapeshifters. Sema Naga wereleopards experienced severe pain and swelling in their knees, elbows, and backs when they shapeshifted, reported Hutton. They were also susceptible to a particular disease of big cats. If a Sema Naga's leopard was being hunted, his relatives stuffed him with ginger to make him more active and his leopard run faster in turn. They could not always run fast enough, however, for several Sema Nagas showed Hutton scars they claimed were the result of bullet wounds suffered by their leopards. Such duplicate wounds did not appear instantly but — conveniently enough — after a few days. If the leopard's wound was fatal, the man also died, but again, in a classic illustration of the power of suggestion, only after hearing the news, which could be several days later. Hutton recalled a man dying on July 19, 1916 following the fatal shooting of his leopard on June 30 the same year.[38]

Some village elders once asked Hutton for permission to tie up a man while they hunted a leopard that had been harassing their livestock. The man himself appeared before Hutton to protest. He was very sorry he was a wereleopard, he said, he did not want to be one, but if he was tied up his leopard would surely be killed, and then he too would die, in effect making it murder. Solomon-like, Hutton gave the elders permission to proceed, but warned that if the man died, then whoever speared the leopard would be tried for murder, and they themselves tried for aiding and abetting the same. The elders decided not to pursue the matter after all. Wrote Hutton: "I was very sorry for this, though I had foreseen it, as it would have been an interesting experiment."[39]

Meanwhile, so intertwined are the shadow souls of Naga shamans with those of their tiger familiars, reported Austrian ethnologist Christoph von Fürer-Haimendorf in 1946, that to all intents and purposes they are one and the same, and so are subject to wound doubling: if either is hurt, so is the other, and when one dies, so does the other. Endicott likewise heard that when a Batek shaman dies, so does his tiger body, and vice versa.[40]

In shapeshifting lore, wound doubling is but one of various motifs that feature the convenient transfer of easily identifiable characteristics, which makes them neat narrative devices and useful "evidence" when you want to incriminate someone, or when you claim to be a shapeshifter yourself. Another is what might be termed diet doubling: that is, any blood sucked or flesh gobbled by a werebeast in its animal form ends up in its human belly (and again, presumably, vice versa). "Evidence" of this is when the "suspect" gains weight, or is supposedly seen vomiting up blood or recognizable pieces of a victim's body, clothing, or jewelry. Again, diet doubling applies whether the shapeshift is physical or spiritual. The Sema Naga man whom the elders wanted tied up told Hutton that if his leopard did not eat, then he too would starve. Hutton also heard that any man whose leopard ate an animal in the night was likely to have bits of its flesh stuck between his teeth in the morning.[41]

Worldwide, a variety of human or humanlike physical characteristics are said to identify a werebeast in its animal form, and weretigers and wereleopards are no exception. The Sema Nagas, noted Hutton, said their mavis had five toes, like a person, instead of the usual four possessed by a cat, often identifying a real leopard's dew-claws, on the forefeet, as the extra digits in question, while in his *Kuang-i chi*, or *Great Book of Marvels*, eighth-century Chinese scholar Tai Fu told of a village headman called Fan Tuan in what is now Sichuan who turned into a tiger that retained his left foot, still booted. Similarly, when Ch'en Shih-san's sister-in-law caught him red-handed, she lashed at him with a flail as he dived under his tiger skin, hitting one of his hands, which he then retained permanently in his tiger form: which is how people were able to recognize him thereafter.[42]

Werecarnivora of all sorts are often said to be tailless in animal form. There is a certain logic to this in that all carnivores have tails but humans do not. In China, wrote de Groot, citing a twelfth-century report, tigers can become men, but when they do their tails remain, and have to be burnt off. Douglas noted that in the last of the infamous late sixteenth-century French werewolf trials, that of 14-year-old Jean Grenier, 13-year-old witness Marguerite Poirier said she knew the wolf she beat off with her staff while tending sheep was Jean because it was smaller and stouter than a real wolf, with a smaller head, red hair like Jean, and a stumpy tail. The man-eating leopard and supposed werecat Burton hunted in Berar in 1891 was widely said by survivors of its attacks to be tailless, not to mention black. (When it was finally killed by villagers, after Burton had wounded it, it was neither.) According to Wessing, Tapanuli Selatan tjindaks are likewise said to be tailless in tiger form: and to have human teeth, bar two fangs in the upper jaw.[43]

As we shall see, however, growing a tail is a key part of the transformation in many stories from the Malay world.

When people wish to "prove" someone has become an animal postmortem, meanwhile, they also sometimes make a connection to the human form. In 1922, Malay scholar Zainul-Abidin bin Ahmad reported that in

Negeri Sembilan there were said to be whole families who post-mortem became tiger-form familiars with the same features they had as people. Wessing noted that Mak Sum's mother was said to have become a five-toed tigress, and cited a story of a Sumatran man with a limp who turned post-mortem into a tiger that limped on its equivalent hindleg. Similarly, heard Endicott, Batek tiger-form ancestors have human facial features, and a Batek shaman's tiger body has his facial features.[44]

I suspect such notions originate from the fact that every tiger has unique face and body markings, making individual tigers readily recognizable.

Presumably the bangles and earrings sported by Fuchs's olthwa enabled those who saw it to identify it as someone they once knew. Finding earrings on an animal is certainly a staple motif in shapeshifting lore. So is following or tracking an animal, then finding only a person. At Fort de Kock, Bradley heard about a man who lived apart from his fellow villagers, kept himself to himself, did little work in the rice fields—but never seemed to be without—and supposedly had no enemies. One evening a tiger pounced on a girl drawing water from the local spring. Hearing her screams, her father hurried over and hurled his spear at the tiger, skewering it in the chest. Alas, he was too late to save her, and the tiger fled. The next morning the villagers followed the bloody tracks of the wounded tiger all the way to the loner's hut—which in typical Malay fashion was built on stilts—and up the ladder. Inside, the man lay dead on the floor, the spear in his chest. "Always, when dying," said Bradley's informant, "the *tjindak* takes on its human form."[45]

Other Sumatran Malays told Bradley tjindaks could only be killed with a silver bullet, or by some other kind of bullet with special powers; that spears and knives glanced harmlessly off them. Silver symbolizing purity, the same is often said of European werewolves, despite Douglas dismissing the idea as a piece of spurious folklore. In colonial India, noted Crooke, silver and gold were both considered to be obnoxious to evil spirits because of their value, scarcity, and brightness. And Locke recalled Malays in the Ibok area of Kemaman telling him that a well-known old tiger there—one that was accorded the title datok and reputedly had all his food fetched for him by younger tigers—could only be killed by a bullet of solid silver. And yet for all this, werebeasts generally are often said to covet their victims' jewelry, and Kerinci weretigers, for one, were said to covet gold above all else, as we shall see."[46]

Prejudice being rooted in fear and ignorance, favorite targets of accusations of sorcery in general and shapeshifting in particular are unpopular individuals, outsiders, and other minorities: anyone different from the accepted social norm. In sixteenth-century France, those accused of being werewolves included a mentally defective beggar (Jacques Roulet), a stunted child who was also an imbecile and a beggar (Grenier), and a curmudgeonly hermit (Gilles Garnier).[47]

Things were no different in Asia. That weretigers in the Malay world are said to lack a philtrum, have crooked feet, and be unable to look you in the eye implies, as Boomgaard notes, that the physically disfigured there were targets of suspicion.[48]

Bradley reckoned the residents of every isolated jungle-edge Sumatran village would say they had at least one willing tjindak living among them. Any man unlucky enough to be branded one had a bad time of it. Such a man in Java, wrote Banner, lives a joyless, friendless life. At best he is shunned, and his wife and children hardly dare show their faces. At worst he is the target of mounting suspicion when a cattle-lifter or man-eater is at large, and soon "there are a dozen who can swear they have peeped into his hut at night and seen a heap of human bones beneath his bed."[49]

De Groot asked of China in days gone by, "Who shall count the hapless men who, suspected of being tigers in disguise, have fallen victims to fear and exasperation? And how many times has this vulgar credulity been aroused against objects of hatred, in order to get them lynched?" Ethnic minorities were common targets of suspicion. De Groot noted that the Man people of Hubei, for one, were said to be weretigers, citing the story of a Jiangxi man who purchased the necessary magic—"a piece of paper with a large tiger painted on it, and a charm beside the beast"—from a Man sorcerer "for three feet of linen, some pecks of rice, a red cock, and a pint of liquor."[50]

In India, as we shall see, all sorts of Untouchables—a group that includes tribespeople—and low-caste Hindus were said to be black magicians, as were wandering Hindu and Muslim ascetics (sadhus and fakirs). Tribespeople everywhere in Asia have long been accused of sorcery. To take but one example, villagers in northern Thailand saw the Mlabri as sorcerers, noted Bernatzik.[51]

In an early example of the Eldorado Syndrome, in the fifth century B.C. the Greek historian Herodotus wrote that the Scythians said of their neighbors the Neuri that they annually turned into wolves for a few days. Such talk—that it is always the people of the next village or tribe who are shapeshifters, or have some other special power—is common worldwide.

"Like all Nagas," wrote Hutton, "the Angamis believe in some village away to the East peopled solely by lycanthropists." Bernatzik noted that when a tiger killed a Mlabri, they in turn blamed their neighbors, the Lao and the Yao.[52]

Blank-faced denial is not always the rule. According to various Dutch reports, the village of Lamongan near Probolinggo in East Java was said to be inhabited solely by hereditary, involuntary spiritual weretigers who would never let outsiders stay the night for fear of what they might do to them.[53]

Meanwhile, Hazeu reported that all the residents of Prata in the Jepara region of Central Java were supposed to be somersault tigers who, in their animal form, raided other villages' livestock: a reputation they cultivated by making any outsider who wished to wed a Prata girl pay a "tribute" to keep them from nightly attacking his own village.[54]

Similarly, various reports told how, in the Lodoyo forest near Blitar in East Java—a forest steeped in weretiger tales—it was said that all the residents of a village called, aptly enough, Gadungan, were somersault tigers who openly threatened people with transforming if they did not give them alms.[55]

Indeed, beggars from all over the Lodoyo area, sometimes wearing tiger skins for effect, wandered around Java knocking on doors and threatening to come back later in tiger form if the occupants did not buy their wares or give them money.[56]

Boomgaard argues that the absence of large-scale witch-hunts against Javanese beggars and other supposed tigermen in Indonesia—including Kerinci pedlars—suggests that the fear they inspired was greater than the resentment they caused. That said, he concedes the occasional lynching did take place.[57] For sure, there is more than a little evidence of the lynching of Kerinci pedlars in the Malay Peninsula, as we shall see.

Tales of women physically transforming into tigers seem to be fairly common only in China. One day in northern Guangdong in the sixth century, wrote de Groot, a young woman called Siao-chu, or Little Pearl, accompanied her brother's wife into the mountains to gather seeds. On the way back they passed a temple, and Little Pearl was so drawn to it she refused to go home. A few nights later, her rain-soaked fiancé, a man called Li Siao, and a comrade were passing when they spied the glow of a fire, so they went inside to dry off and get warm. On the altar they noticed a pile of clothes. Hearing footsteps, they hid behind a screen; and in walked a tiger. To their amazement, it "divested itself of its teeth and claws, rolled up its skin," placed them on the altar, and put on the clothes, at which they saw it was Little Pearl. Li Siao embraced her, but she did not speak. At daybreak he took her home, where her family locked her in an outhouse and threw her raw meat, which she readily devoured. But they noticed she kept staring at the pig, and when, a few days later, she changed again into a tiger, the villagers all armed themselves with bows and shot her dead. (She gained her revenge post-mortem, though, ravaging the area as a terrible man-eater for a whole year.)[58]

De Groot also told of a poor man called Yuen Siang two centuries earlier whose luck appeared to change when he was on his way home one evening and met a young woman "whose charms and beauties were perfect in every respect." Before long they were married and had twin boys, and by the time the lads were ten Yuen Siang was a wealthy man. But one night his wife was seen hurrying to the grave of a just-buried villager, taking off her clothes and hair pins and hanging them on a tree, then changing into a tiger, in which form, just like a tjindak, she disinterred and devoured the corpse before resuming human form. From that night on she "scoured the district and its hills, devouring corpses again and again."[59]

In another de Groot story, the villainous "woman" is not actually human, but a demon in disguise. A man called Ts'ui T'ao stops at an inn one night despite its keeper urging him not to. T'ao is just about to go to sleep when a tiger walks in. Luckily he has time to hide, and from his hiding place he sees it "put off its animal skin in the middle of the courtyard and become a girl of extraordinary beauty, well dressed and ornamented," who promptly makes herself comfortable on his bed. T'ao confronts her, but she convinces him he merely saw her take off a tiger-skin dress, and tells him straight up she is a poor girl in need of a husband. T'ao instantly falls for her charms, throws the "dress" down the innkeeper's well, and takes her as his wife. Fast-forward a few years, and when the happy couple and their children happen to stop one night at the same inn, T'ao is amused to see the "dress" still down the well. "Have it fetched up, and let me try it on again," says his "wife," so he does. She turns straight into a tiger and eats up both T'ao and the children.[60]

Not all alluring weretigresses in China devour their men. De Groot also told of a man called Sü Hwan who, one day in A.D. 396, met a maid "who fascinated him by a ditty, and wanted him to go between the shrubs." There she turned into a tigress and carried him on her back off into the hills, before returning him home — no doubt exhausted — ten days later.[61]

Thomas Aylesworth, author of various sensationalist books, wrote in 1970 that in Cambodia a woman can be physically turned into a tiger by smearing her with a magic ointment, after which she runs into the forest and transforms within seven days. To reverse the transformation, a man rubbed with the same ointment tracks her down and hits her over the head with a club.[62]

More sensibly, Bradley heard tell at Fort de Kock of a man who had a beautiful and devoted wife he adored, even though she was always sleepy by day and often stole out at night, making all sorts of excuses when she returned. One night he followed her and saw her head off into the jungle. When he confronted her about this the next day she merely said that she had felt ill and needed to bathe in the river. But as the days passed she ate less and less rice and more and more meat; almost raw meat, at that. Then, one evening, a tiger killed a girl from the village, and when the man ate his dinner that night he found a long black hair in his meat. It could not have been one of his wife's, since her hair was cropped. She lamely said that she had combed a friend's hair on her way home. That night she slept soundly, with blood on her lips, but the next day, when she saw that he could not look her in the eye, she left home for good.[63]

Bradley also heard that when the Chinese owner of a rubber plantation at somewhere called Labocon Bihil once set a gin-trap for a tiger, he instead caught one of his female employees. Concluding she was a tjindak, he and most of her co-workers left her to her fate, but luckily some friends chanced by and set her free. Though badly hurt, she bore those who had ignored her cries for help no ill will, thinking it perfectly natural.[64]

Variants of this story seem to have been common wherever people set traps for tigers, but usually the person caught is a man. One, with a very different outcome, was a stock story in the Malay Peninsula, as we shall see, while another, cited by de Groot, told how some villagers in the mountains of Hunan once set a trap — one to catch tigers alive, this time — and caught instead a stranger, an official of some sort. On asking him how he came to be inside, he replied angrily, "The prefect of this district sent for me yesterday; but I had to skulk somewhere in the dark against the rain, and thus inadvertently got into this trap; be quick and let me out." But the villagers demanded to see proof, and only let him go when he produced the official summons. He immediately turned into a tiger and ran off.[65]

Outside China, most yarns about physically transforming weretigers (and wereleopards) are about men — just as almost all the werewolves of medieval Europe were said to be men — and when a woman does feature in one it is more often in the role of a wife who in some way or other brings about her husband's downfall, or who is an unsuspecting "lion bride," a woman who discovers too late her husband's true nature.

In 1890, Georgiana Kingscote and Pandit Natesa Sastri recorded a tale from southern India that not only is a classic "lion bride" tale, as well as the origin of an old Tamil proverb, but features a tiger that turns into a man, not vice versa. In a house in the forest there lived a tiger who had acquired great proficiency in magic. Fond of the occasional vegetarian meal, he would assume the shape of an old Brahman, enter a nearby village and share the food prepared for the Brahmans there. But above all he desired a Brahman wife to cook for him at home, so when he heard that there was a pretty young Brahman girl in the village whose parents were anxious to see married, he assumed the form of a handsome and scholarly young Brahman, sat on a rock by the village burning-ghat, scattered ashes over his head, and began reading out aloud from the Ramayana. When the girl came to bathe she instantly fell in love with him. Soon they were married, and for a while all went well. But after a month in his in-laws' house he missed eating meat, so he told them that it was time to take their daughter to live in his own home, which he gave as being in a conveniently distant village.

The walk through the forest was tiring for the girl, but when she said that she wished to rest, her husband scolded her, saying, "Be quiet, or I shall show you my original shape." Scared of him for the first time, she went on in silence, until at last she could bear it no more and asked again if they might stop. When he answered as before she angrily defied him to do as he threatened, and to her horror he turned into a tiger. Only his voice remained unchanged, as he chillingly told her that he would kill her if she didn't live in his house and cook for him every day. Weeping, she followed him to his home.

Every day he brought home meat and vegetables for her to cook, though he took his meals outside and seldom went indoors. A while later she bore him a son; a tiger cub, conceived when he was in human form. And so the years passed, until one day a crow saw her crying, asked her what was wrong and, on hearing her tale, agreed to help her. With an iron nail she wrote a letter on a palmyra leaf, which the crow took straight to her three brothers. The young men set out at once and on the way collected a donkey, an ant, a palmyra tree and an iron tub.

On their arrival their sister hid them in the loft, as her husband was due home any minute. When he returned he became suspicious and demanded to know who was inside with his wife. She called out that she was alone, but just then the brothers saw him and fell out of the loft with a crash. She then admitted they were there and said they wished to speak to him after he had eaten. The tiger demanded to hear their voices, so the youngest brother put the ant in the donkey's ear, and on being stung it started braying. Alarmed, the tiger demanded to see their legs, so the eldest brother stretched out the palmyra tree. Finally the tiger demanded to see their bellies, so the middle brother held out the iron tub. Now thoroughly frightened, the tiger ran away.

When he returned he found that his wife had torn their son in half and roasted him over the fire then fled with her brothers and as many of his possessions as they could carry. Bent on revenge, he went to her home in human form. There he was greeted warmly and invited to sit on a seat of sticks. These concealed an old well and when he sat down he duly plunged to his death. The brothers sealed up the well, but the girl raised a pillar over it — representing Shiva — and planted a sacred Tulsi, or Holy Basil plant, on top. And every day she smeared sacred cow dung on the pillar and watered the herb, in memory of her husband.[66]

In Hindu mythology the palmyra — *Borassus flabellifer*, a tall fan-palm — is the living descendant of the original *kulpa briksha*: the wish-giving tree, or tree of eternal life, and one of fourteen remarkable things revealed when the gods and demons stirred up the ocean. The original kulpa briksha now stands in the first heaven, but the names of Rama and his wife, Sita — incarnations of the benevolent deities Vishnu, preserver of the cosmos, and his wife, Lakshmi — are written on the trunks of all its descendants on Earth. Anyone who touches one is therefore safe from any animal. Palmyra has also long been associated with the healing power of tigers. In 1916, Frederik van Ossenbruggen wrote that wooden amulets in the shape of tigers and decorated with palmyra leaves, found on the Indonesian island of Nias, were said to draw illness out of the wearer.[67]

But the tale of the Brahman girl who married a tiger is notable above all for its portrayal of the tiger as a cunning yet easily deceived creature that lives in limbo between the forest world, where it really belongs, and the human world, to which it is drawn but to which it can never properly belong: a world from which, many would say, it originally came, but to which it can never return. Just as no human can ever really be a tiger, then, so no tiger can ever truly be human.

6

Tigermen in Malaya: Negritos and Jambi Men Accused

> "All these things the Malays know have happened, and are happening to-day, in the land in which they live, and with these plain evidences before their eyes, the empty assurances of the enlightened European that Were-Tigers do not, and never did exist, excite derision not unmingled with contempt."—Hugh Clifford, In Court and Kampong, 1897, 65–6

Born into one of England's leading landed families, Hugh Clifford had a sheltered upbringing in the West Country, being educated at home. In 1883, aged seventeen, he won a place at Sandhurst but, drawn like Hicks, perhaps, by the romance of the East, instead of following his father into the army he joined the Civil Service of the Protected Malay States of the Malay Peninsula — Perak, Negeri Sembilan, and Selangor — of which an uncle, Frederick Weld, was high commissioner. During the next twenty years he spent long periods living and working in remote areas of the peninsula, particularly Pahang, where in 1887 he became that state's first British Agent when still only twenty.[1]

In 1903, Clifford was made colonial secretary of Trinidad and Tobago, and subsequently he was governor of Ceylon (now Sri Lanka), the Gold Coast (now Ghana), Nigeria, and Ceylon again, being knighted in 1909. But the Malay Peninsula was his true love, and in 1927 he returned in triumph as governor of the Straits Settlements and high commissioner of the Federated Malay States — which by then included Kelantan, Kedah, Perlis, Terengganu, and Johor — where he made nostalgic visits to all his old stamping grounds.

One day in 1928, during his first few sweltering weeks in the East, a young civil servant called Mervyn Sheppard was out in his district in Pahang collecting rent from an isolated kampong when he and the penghulu were amazed to hear an automobile approaching, for there were scarcely any motor vehicles in the peninsula at the time. It screeched to a halt, and out stepped a tall, stooping figure in a sola topi and full-length, high-collared white uniform: Sir Hugh Clifford. Recognizing him at once, after nearly forty years, the penghulu ran through the gathering crowd to greet him. The old friends touched hands and salaamed, then Clifford made a short speech, in both Malay and Pahang patois, that Sheppard could not follow. "But everyone else was engrossed," he remembered, "enthralled by what he was saying. Then at the end of his brief address he turned round, got back into his car and was driven away."[2]

Sadly for Clifford, the changes he found on his return destroyed the idealized vision of rural Malay life he had clung to in his long years away. Disillusioned, his mental health deteriorated rapidly, and he had to retire after only two years, by which time he was known fondly by fellow colonials as "Mad Clifford."[3]

Peninsular Malays, wrote Clifford, did not just *believe* in weretigers, they *knew* they existed. Specifically, they knew that "the worst and most rapacious of the man-eaters are themselves human beings, who have been driven to temporarily assume the form of an animal, by the aid of Black Art, in order to satisfy their overpowering lust for blood." Even in the 1950s, noted Locke, this was still "believed implicitly, not only by the illiterate village Malays and Chinese labouring classes, but by many educated members of the two races."[4]

In 1897, in a dramatized account "more or less founded on fact," as he put it in his Preface, Clifford recalled a tragic event that reputedly took place in a kampong called Ranggul on the Tembeling River in Pahang in the early 1860s. Nine Malays were gathered one evening in a large, isolated house at one end of the village: the owner Che Seman, his wife Iang, their daughter Minah and sons Awang and Ngah, their cousin Abdollah, a relative called Mat, and two friends, Potek and Kassim. The men sat on the floor eating curry and rice with their fingers, while the women knelt to one side, ministering to their needs. In Clifford's words: "There was no sound to be heard, save the hum of the insects out of doors, the deep note of the bull-frogs in the rice swamps, and the unnecessarily loud noise of mastication made by the men as they ate."

When the men had eaten their fill they chewed betel leaves and rolled cigarettes, while the women ate what was left, and at last broke their silence. Abdollah told how his wife's aunt's brother had fallen to his death from a coconut tree, and Mat added a few unnecessary details about the corpse. Potek and Kassim talked wildly of the army the Sultan of Pahang was gathering, of his thousands of elephants and tens of thousands of followers. But before long Che Seman raised the subject uppermost in everyone's minds, causing them all to fall silent, even the women. "He of the Hairy Face" was with them once more. Only the previous evening he, Che Seman, had met a local imam and invited him to supper, but the prayer-leader had declined the invitation because the man-eater had struck again the night before and it was best to be home by nightfall.

Awang whispered fearfully that this took its tally of victims to twenty-three in three months, adding that it killed neither goats nor cattle, and that people were saying "He sucketh more blood than He eateth flesh."

That, said Ngah, is what proved him to be a weretiger.

Che Seman solemnly agreed. The man-eater was a Negrito cast out by his own kind. By day he went about in human guise. He, Che Seman, among many others, had seen him roaming naked in the forest, muttering to himself. But at night he assumed the form of a tiger and kept constant watch for anyone foolish enough to venture outdoors. "He cometh like a shadow," said Che Seman, "and slays like a prince, and then like a shadow he is gone! And the tale of his kills waxes even longer and yet more long. May God send Him far from us!"

But at that very moment they heard the tiger's distant roar. In silent terror they listened as the tiger roared again and again, ever louder and nearer. Finally it roared within the very grounds of the house, startling Mat into knocking over and extinguishing their only torch. The women huddled against the men, who gripped their spears and trembled in the dim glow cast by the dying embers of the fire and the faint moonbeams creeping through gaps in the roof. "Fear nothing, Minah," whispered Che Seman, "in a space He will be gone."

But seconds later everyone screamed as the tiger leapt against the side of the house and fell back to the ground with a roar. The men tried feebly to rally their spirits with a *sorak*, or Malayan war-cry; all, that is, bar Mat, who crept away and hid in an upper compartment. After a minute or two of awful suspense the tiger again leapt against the house, and this time succeeded in clawing its way up onto the roof. The sorak died in the men's throats, and all watched in terrible, silent fascination as the tiger tore through the thatch. Momentarily they saw its moonlit face framed in the hole, then it was among them.

The tiger struck rapidly, crushing skull after skull with great swipes of its paws. Soon all were dead except for Mat, who cowered, watching, overhead, and Minah who, though badly wounded, managed to crawl under her father's body. Purring contentedly, the tiger drank deep of each victim's blood, until at last it found the girl. There followed a ghastly scene, the tiger toying with her like a cat with a mouse — for as long, said Mat later, as it takes to cook rice — before finally ending her misery (Figure 1, in the introduction). Still it was not finished, and Mat watched trembling as it played with each of the mangled bodies in turn throughout the long night, mutilating them repeatedly with its teeth and claws. Only at daybreak did it abandon its dreadful game, break down the door, and make off into the jungle. When neighbors arrived later they found a scene of carnage, with Mat still cowering in his hiding place, half mad with shock.[5]

So far as he knew, noted Clifford, this was the only recorded instance in the peninsula of a tiger attacking anyone in their home, so it was little wonder Malays believed it was a weretiger. Perry speculated that the tiger was rabid, citing two similar cases in Assam — the first in 1943, when a tiger in the Nagaon district attacked eighteen people in their homes, killing eleven, and another in 1950, when a tiger broke into several houses one night and attacked fourteen people, killing three — but thought earlier reports of tigers "running amok" (an appropriately Malaysian simile) may well have involved rabid animals.[6]

Occasionally, a wound is enough to make a tiger "run amok." Shortly before Hicks arrived in the Wardha district of the CPs in 1881, a sportsman shot at a tiger in the Dugger Forests there and hit it in a hindleg. The wound festered and filled with maggots, which drove the tiger literally mad, reckoned Hicks, causing it to rampage around and kill everything in its path. (Hicks eventually shot it dead, but not before it had killed a Gond beater.)[7]

But the simplest explanation for the Ranggul massacre is that the tiger acted instinctively, like a fox in a henhouse. Either that, or it was the work of human hands, even someone — Mat, say? — run amok: though I would be the last to suggest anything so cynical.

The Batek told Endicott that while in theory one of their shamans could use his tiger body to kill someone — his mother-in-law, for instance — in practice this was almost unheard of. One exception was a shaman who lived on the upper Tahan River but used his tiger body to go on a killing spree in far off Perak: when his tiger body was shot dead in Perak, he too subsequently died. That said, wrote Endicott, over the years the Batek had been happy to perpetuate the fear among Malays that some of their number were weretigers, and physical shapeshifters at that, because it gained them respect.[8]

This was a view shared some seventy years earlier by Skeat and Blagden, who reckoned the various tribespeople of the Malay Peninsula started the rumors they were weretigers themselves, "with the object of impressing their more civilized neighbours with all the fear they could, an object in which they obtained a considerable measure of success." On his 1899–1900 Cambridge Expedition with Blagden, at Ulu Aring, a little village at the headwaters, or Ulu, of the Aring River, Skeat met a Negrito shaman called Pandak—undoubtedly a Batek, reckoned Endicott—who had a great reputation among his fellow villagers as a dangerous weretiger: a reputation Pandak himself was happy to promote.

When he wanted to become a tiger, Pandak said, he announced he was going for a walk. All the time he was away, though, it was essential his wife kept a fire burning and occasionally lit incense, otherwise he would never return. Deep in the forest he lit some incense himself, trapped some of the smoke with his right hand, held the clenched fist to his mouth, and blew through it over each shoulder, then straight ahead, all the time invoking the spirits of the mountains to grant him his wish. Squatting, he blew more smoke through his fist, leant forward, and shook his head from side to side. At this he sprouted fur, grew a tail, and turned into a tiger. In this guise, for the next seven to twelve days, wrote Skeat, "he would feast upon bodies of his victims (whether dead or alive), always, however, excepting and burying the heads," but whether these victims were human or animal is not clear. What is certain is that during his time as a tiger he raided cattle-pens until his craving for raw beef was assuaged. Man-eater or live-stock-lifter, when his time as a tiger was up, Pandak returned to the same spot, squatted, and said, "I am going home." On regaining his human form he would vomit up all the undigested bones he had eaten as a tiger.[9]

Tribespeople living deep in their own forests are not the only ones peninsular Malays have traditionally said are weretigers. They and Sumatran Malays have long also said it of Kerinci men, many of whom once traveled around both Sumatra and the Malay Peninsula as pedlars. Given the claims and antics of Kerinci dukuns, this is perhaps hardly surprising. Boomgaard noted there were Kerinci traders from at least as early as the seventeenth century. Originally there may have been a few wealthy ones, trading in forest products and gold, and they certainly acquired a reputation as misers, but as their numbers increased they became poorer and roamed further and further afield in search of gainful employment— Bradley referred to them as "those splendid carriers that bear the coffee to the coast on their backs in sixty-pound loads"—with many becoming little more than wandering beggars and petty criminals. That such vagrants might possess gold only added to their supernatural reputations. When Skeat once asked some Malays at Jugra in Selangor how it could be proved that Kerinci men were weretigers, they told him a Kerinci man with gold teeth was once killed in his tiger form, and on being examined the tiger was found to have matching gold teeth (which is perhaps taking the transfer of an easily identifiable characteristic a tad far).[10]

Another typical trait of supposed Kerinci weretigers was that they were pious ascetics who had made the long haj, or pilgrimage, to Mecca, something that was thought to greatly increase the powers of Malay *kiais*, or holy men. Take the case of the celebrations once held by Javanese noblemen. These began with a tiger (nature, the uncouth Dutch) and a water buffalo (cultivation, Javanese nobility) being made to fight to the death — the latter usually won — and ended with a *rampok macan*, at which a tiger was released into a ring of spearmen. Wessing noted that, according to some reports at least, the tiger was released by a kiai, who on opening its cage calmly turned his back and walked away. The kiai's power over animals was attributed to him having made the pilgrimage to Mecca. Moreover, he was said to be able to turn himself into a tiger.[11]

Boomgaard reckons that, like itinerant beggars from Lodoyo in Java, at one time at least Kerinci pedlars probably exploited, even cultivated, their reputation by threatening people with transforming if they did not buy their goods or give them alms, particularly gold: that like the Javanese beggars they could get away with this without inciting witch-hunts because the fear they inspired was greater than the resentment they caused.[12]

However, in no story I know of from the Malay Peninsula — stories written from around 1900 onwards— do Kerinci men threaten people with their supposed powers. On the contrary, they are typically portrayed as law-abiding citizens who strongly protest their innocence before being lynched.

Unsurprisingly, British colonials in the Malay Peninsula were far from convinced Kerincis were weretigers. Typically sceptical was Keyser, who recalled camping one night in a remote area of jungle — feeling secure inside his mosquito-net, which he reckoned any prowling tigers would think was a trap and so give a wide berth — when he noticed some of his Malay followers keeping watch when usually they would have been fast asleep. When he asked them why, they were indignant. Surely the Tuan knew there were tribes in the area who kept tigers as pets and sent them out at night to fetch meat? And not only that. They had been too frightened to mention it before, but just a few nights earlier they had seen two colleagues—"Jambi men," as they put it—slip away, return at dawn as tigers, and change back into men, with blood dripping from their mouths, as they lay down again. "I thanked them for their vigilance," wrote Keyser, "and turned over to resume my slumbers."[13]

Another sceptic was Keyser's boss, Frank Swettenham, a main player in bringing the whole peninsula under British control. Born into a family of colonial administration, he joined the Civil Service of the Straits Settlements—Penang, Singapore, and Malacca—as an ambitious 20-year-old cadet in 1871, and rose rapidly, being knighted in 1897, to become both governor of the Straits Settlements and high commissioner of the by then Federated Malay States of Perak, Negeri Sembilan, Selangor, and Pahang in 1901.[14]

Swettenham was a very different man than Clifford, with a very different perspective on Malay life. While the courtly, reserved Clifford was plagued by doubts about his role, the autocratic Swettenham was absolutely sure of his. Malays, he reckoned, were a lazy, childlike, backward people, "passed by in the race for civilization," who would be much better off under British rule than under the sultans. British rule would bring them "regeneration," but until then they would remain shackled by feudalism and superstition.[15]

All the same, as with Clifford, Malays held Swettenham in the highest regard. A cholera epidemic once broke out in a district, and he went to see how he could help. He found the usual scare absent and everyone strangely calm. It transpired a pawang had turned up just after the outbreak, sold 500 people a string bracelet at a dollar a piece, then left in a hurry. Swettenham told some of them they had been conned, that the charm was useless. "But we told him so," they countered, "and he promised that if any one who bought the charm was attacked by cholera, and died, he would, in every such case, give back the dollar." (Leopold Ainsworth, a rubber-planter, told exactly the same story more than thirty years later, only casting himself as the enlightened colonial concerned.)[16]

Almost all peninsular Malays, noted Swettenham in 1895, believed that some Kerinci men could turn themselves into tigers and "in that disguise ... wreak vengeance on those they wish to injure." But it was only fair to point out, he went on, that the Kerinci vehemently denied this, saying the real culprits were "the inhabitants of a district called Chenaku in the interior of the Korinchi country," and subsequent writers have repeated Swettenham's words in good faith.[17]

The trouble is, it seems there never was such a district. There is a tributary of the Indragiri River in Riau province in central Sumatra, the Tjenako, which was spelt "Chenaku" by nineteenth-century writers—civil engineer James Motley, for one — but most likely Swettenham got his wires crossed. The first mention of Kerinci weretigers was by Arend van Hasselt, who explored Sumatra in the 1870s, and wrote in 1882 of hearing about a district "to the side of Kerinci" called "Banije-balingka" which had two villages populated by "oerang Tjindakoe," or Tjindakoe people.[18]

Here, then, is the origin of the word "tjindaku" or "tjindak." This is a connection writers such as Boomgaard and Wessing surprisingly do not comment on when they transcribe "oerang Tjindakoe" as "the Cindaku people."[19]

As far as I can ascertain, there never was such a people as "the Cindaku," and they are entirely mythical.

Anyway, in one village, wrote van Hasselt, the Tjindakoe were boarmen; in the other, tigermen. At certain times they all set out in their animal form, the boarmen to raid farmers' crops, and the tigermen in search of human prey. When the tigermen reached a river too wide to swim across they assumed human form, disguised themselves as merchants carrying trading packs, and asked to be ferried across. On reaching a village they sought shelter for the night, then when their hosts were asleep turned back into tigers, killed them, and ate their hearts: just like the tjindaks of Tapanuli Selatan. Meanwhile their king stayed behind, chained by his navel to a rock. On their return they presented him with a bitter-tasting, heart-shaped banana flower. If they brought him a real human heart, its sweet taste would drive him to break free and rampage around looking for more.[20]

The Kerinci themselves told Bakels the tigermen village had nothing to do with them and was actually, as Marsden reported in 1784, in the Pasemah area of Bengkulu, south of Rejang: near Mount Dempo, in fact. In the same area in 1817, Edward Presgrave, who served under Raffles in Bengkulu, heard it said that somewhere in the locality was a stream that turned you into a tiger when you crossed it, and reversed the transformation when you crossed back, so perhaps the Pasemah tjindaks used that. ("From this fabulous story we expected to find the woods infested with tigers," commented Presgrave, "but to our astonishment we discovered nothing.") The Kerinci told Bakels the king sends his subjects out for four months at a time each year to fetch him human hearts. Most Kerinci were terrified of these tjindaks, and depended on their dukuns to keep them away. Some, though, reckoned they only harmed you if you broke adat rules, showing how ideas about tiger-form ancestors and weretigers can merge.[21]

Wrote Swettenham: "At night when respectable members of society should be in bed, the Korinchi man slips down from his hut, and, assuming the form of a tiger, goes about 'seeking whom he may devour.'" But the cases he cites relate only to the loss of poultry. Once, four Kerinci men arrived in a district in Perak, and a tiger killed and ate several chickens in the area that same night. They left the district, but a short while later three of them returned, said a tiger had just been killed, and begged the local penghulu to give it a decent burial. On another occasion, some Kerinci men sought shelter in a Malay house: that night a tiger took some chickens from the kampong, and the next day one of them vomited chicken feathers.[22]

Clifford, too, wrote that all Malays knew of "the countless Korinchi men who have vomited feathers, after feasting upon fowls, when for the nonce [occasion] they had assumed the form of tigers." And in another account "more or less founded on fact," Mat Saleh, an old penghulu of a kampong in the Selim valley in Perak, supposedly once told him the following story, which he swore was true. Into the valley one day came a wealthy Kerinci pedlar, Haji Ali, with his two sons, Abdulrahman and Abas. All three were weighed down with heavy packs of sarongs, which they proceeded to hawk with much crafty haggling and hard bargaining. Haji Ali liked the place so much, he bought some land to farm, and settled in a fine new house.

They were quiet and respectable, attended Friday prayers and, being wealthy, soon found favor with their poorer neighbors: even more so when Haji Ali, a widower, let it be known he wished to remarry. There was much excitement among those with daughters of a marriageable age and, though he was well past his prime, he had his pick. Eventually he chose a pretty girl called Patimah, who jumped at the chance to escape a life of poverty and drudgery. Yet only three days later she was back, pounding on her parents' door at dawn, trembling, disheveled, and terrified.

Patimah told a remarkable tale. Her husband and his sons had treated her well, and even when she cooked the rice badly and the sons grumbled, Haji Ali did not blame her, when she had expected a thrashing. Nevertheless, she declared, nothing her parents could do or say would make her go back to a man who hunted at night as a weretiger.

Each evening after prayers, she said, her husband had gone out on some pretext or other, not returning until just before dawn. On the third night, when she heard him coming back, she hastened to open the door, and saw on the ladder down to the ground, resting between two paws on the top rung, the head of a fully grown tiger, with "long cruel teeth," and a "fierce green light" in its eyes. Below the head the creature's stripes showed boldly, but the lower half of its body was that of a man (Figure 10). Patimah stared, too terrified even to scream. Then, wrote Clifford: "Slowly, as one sees a ripple of wind pass over the surface of still water, the tiger's features palpitated and were changed, until the horrified girl saw the face of her husband come up through that of the beast." Regaining her senses, she leapt past him and fled into the jungle.

Patimah's story soon got around and no one questioned it, for what else could have driven her into the jungle at such an hour? Indeed, the disappointed parents of other girls of a marriageable age took great delight in telling her mother and father they had expected something of the sort to happen all along. Conclusively, Haji Ali made no attempt to get his young wife back. Everyone shunned him and his sons thereafter.

That was far from being the end of the matter, however. One night a tiger killed a buffalo belonging to Mat Saleh. Telling no one, the penghulu set a trip-wire gun-trap over the carcass. Sure enough, the following night the gun went off.

The next morning Mat Saleh had no difficulty in rounding up a dozen volunteers. Armed with an assortment of guns, spears, and knives, the party sallied forth to check the trap, where they found fresh paw prints and great splashes of blood. The trail was easy to follow, for as well as dripping blood the creature had dragged its right hindleg. All the same, they proceeded cautiously.

The trail led straight to the house of Haji Ali. Inside they found Abas, who surlily told them his father was sick. At this the party grew excited. What was that large patch of blood outside, inquired Mat Saleh casually? Abas said they had slaughtered a goat the night before. When the penghulu asked if he might buy the skin, Abas said they had thrown it in the river because it was mangy. What was wrong with his father, persisted Mat Saleh?

Before Abas could answer, his older brother came through the curtained doorway from the inner room and told the penghulu and his men to clear off, as they were disturbing his father on his sickbed. "He held a sword in his hand," wrote Clifford, "and his face wore an ugly look as his words came harshly and gratingly with the foreign accent of the Korinchi people." Mat Aleh had no choice but to lead his men out. As they descended

6. Tigermen in Malaya

Figure 10. Cult American illustrator Mahlon Blaine's depiction of the Kerinci Haji Ali reverting to human form on returning home from a night's hunting as a weretiger in the classic "lion bride" tale recounted by Hugh Clifford. Wrote Clifford: "The moon was behind a cloud, and the light she cast was dim, but [Haji Ali's Malay wife] Patimah saw clearly enough the sight which had driven her mad with terror." (From the 1927 William Heinemann, London edition of *The Further Side of Silence*).

the ladder one of them drew his attention to what appeared to be a pool of bloody water on the ground under the house, below the inner room. But with both sons in close attendance they had no chance to examine it.

The penghulu set off at once to report the matter to the district officer. But as he expected, the Englishman dismissed his story, and when he returned dissatisfied, five days later, it was to find the birds had flown. They had fled downriver one dark night, abandoning their house and land and all their uncollected debts. This, remarked Clifford, was the strangest circumstance of all, given the passion for wealth and property in the heart of every Kerinci. To the European mind, he conceded, the only possible explanation was that they had been disposed of by the Malay villagers, and the rest was just a ridiculous cover story, and he himself would be inclined to agree with such an astute explanation were it not for the fact that Haji Ali and his sons turned up in another part of the peninsula some months later, with Haji Ali lame in his right leg.[23]

Half a century after Clifford, Locke reported hearing several versions of this story in Terengganu. In one he briefly tells the order of transformation from tiger back to man is reversed, Haji Ali coming up the ladder with his own head and the body of a tiger. "As the horrified girl watched," he wrote, "the body gradually turned into that of a man, the process finally ending with the disappearance of the tail."[24]

Just as de Groot told how some villagers in Hunan once caught an official alive in a tiger trap and let him go, at which the man turned into a tiger, so Clifford noted that it was common knowledge among peninsular Malays that a Kerinci called Haji Abdallah was once caught alive in a drop-door tiger-trap at Sayong in Perak, "and thereafter purchased his liberty at the price of the buffaloes he had slain, while he marauded in the likeness of a beast."[25]

Ten years after Clifford, in a highly fictionalized account, another colonial administrator, William George Maxwell — eldest son of William Edward Maxwell, a bitter rival of Swettenham — offered a version of this story in which even being well known to local Malays was not enough to save a Kerinci from the mob. Writing as George Maxwell, he told how he once traveled inland from the east coast of the peninsula through a state he described as a long, narrow strip — surely Terengganu — with a Malay district datok called Kaya and a few Malay boatmen, on a journey of at least two days, first by boat up one river, then by foot across a jungle-clad ridge to another river, which they then followed downstream in a requisitioned dugout towards a kampong called Bentong. (This he described as a village of some importance, comprising some fifty houses, in a sparsely populated district, but whether he meant Bentong in Pahang, nearer the west coast than the east and now a large town, I do not know.)

As they drifted downstream around bend after forest-clad bend in the river in the full blaze of the sun, Maxwell and Kaya lounged in the shade of the awning and passed the long journey in conversation. Eventually the talk turned to weretigers, which Kaya said was fitting, given that only a few years previously such a creature had terrorized Bentong. After lifting Swettenham's chicken-feather anecdote and putting it into Kaya's mouth, Maxwell then related Kaya's story; though not, he admitted, quite as the chief told it.

A tiger was regularly attacking the villagers' buffaloes, hardly a month going by without it taking one or two. All attempts to trap it had failed, and the villagers were so desperate they had begun to consider abandoning Bentong, for without their buffaloes to plow their paddy fields they would be ruined.

Late one afternoon an old Kerinci cloth-pedlar called Haji Brahim, a man who had plied his trade in the district for some years, was hurrying on foot through the torrential rain and gathering gloom from the neighboring village of Siputeh towards Bentong, where he planned to spend the night. (The only Siputehs I know of are one in Kedah, and one in Perak.) The downpour and slippery path had made him much later than he intended, for he had heard about the tiger and knew that cattle-killers all too often become man-eaters. Even as he was thinking this he heard the tiger roar close by. Quaking with terror, he ran for his life. As he ran he saw a drop-door tiger-trap, baited with a live dog, on the path ahead. Reasoning that what would keep a tiger in would also keep one out, he crawled into the cage, and the heavy door dropped behind him, shutting him safe inside. Here he spent a long, uncomfortable night, but with the tiger prowling and roaring nearby all the while, "he forgave the presence of the unclean dog that cowered beside him, and blessed the thought that had led him to seek such a refuge."

When morning came the Kerinci found himself unable to raise the door, so he sat back and waited for help. In due course he saw a man coming down the path from the kampong, and shouted to attract his attention. At first the man stared at Haji Brahim dim-wittedly. Then he recognized him, but instead of coming to his aid, he screamed and fled back up the path.

Soon the boom of the mosque drum echoed through the jungle, and in answer to its call all the able-bodied men of Bentong hurried to the house of their rajah, Alang, armed with knives and spears. Rajah Alang led the way to the trap and demanded Haji Brahim explain himself. The Kerinci's heart sank. He realized he was on trial for his life.

Haji Brahim explained how he came to be in the trap, but the crowd murmured its disbelief. "The tracks will prove the truth of what I say," he cried. This seemed fair enough, so the ground was carefully examined. But the men had trampled his footprints, and all they could see were paw prints. "The tiger was here last night, and you are in the tiger-trap this morning," said one.

Seeing reasoning was futile, the old Kerinci offered to swear his innocence on the Koran, but the men rejected the idea out of hand, for clearly so unnatural a creature as a weretiger would not hesitate to make such an offer to save its skin. As a last resort Haji Brahim threw himself on their mercy, begging for his life, promising to do anything, even leave the country for ever, if they would only let him go.

Rajah Alang asked his men what they thought. One stepped forward and spoke for all. By Allah's good grace they had caught the creature ruining their livelihoods: it would be madness to set it free. At a signal from the rajah, another man drove his spear through the bars of the trap, deep into Haji Brahim's side.

When Kaya finished telling his version of the story — which naturally started with finding the Kerinci in the trap — Maxwell expressed his pity for Haji Brahim, to which the Malay replied that a tiger showed no pity and could expect none in return. The men of Bentong had known Haji Brahim for many years, and had nothing against him personally. Rajah Alang was a just man, and Haji Brahim had been fairly tried and clearly proven guilty. The rajah could not have acted otherwise. Maxwell retorted that the evidence was circumstantial. The chief agreed, but pointed out that the British had hanged men for less.

"I could not think of a suitable reply..." wrote Maxwell. "There was, therefore, silence for a space as our little boat broke the sparkle of the river." They were passing Bentong, which looked as quiet and peaceful as any other kampong. A canoe headed out towards them, bearing a noble old Malay: Rajah Alang, no less. Concluded Maxwell: "We stopped for a while to exchange the greetings and the courtesies due to, and expected from, our various ranks. Then we parted, and at the next bend of the river the great forest swept down again to the bank on either side."[26]

Several colonial authors after Maxwell offered their own versions of this bleak tale, citing it as one commonly told by peninsular Malays. Like Maxwell, they all start with the man seeking refuge in the trap, and not, as Malays obviously always told it, with him being found inside it.

In a pithy 1932 version by Betty Lumsden Milne, a friend of Locke, the man is not identified as a Kerinci, but is a stranger called Mat Awang. Suspicion falls on him when goats start to disappear soon after he moves into a kampong called Susa and he is regularly seen leaving his house at night and returning at dawn. In fact, wrote Lumsden Milne, he was probably just visiting some girl in a neighboring kampong. When the villagers decide to set a trap for him, baited with a live goat, he even offers to help them build it. After the villagers spear him to death, one of them remarks, "It was strange that he did not eat the goat."[27]

Following Lumsden Milne, a 1935 version by Redmond Burke (a district officer in India) and his daughter Norah, and another version four years later by J.B.H. Thurston, both promote the tiger to man-eater. In the former the Kerinci is peddling brass, is a stranger passing through the Bentong area, and is only speared after he rashly identifies himself as a Kerinci. In the latter the leader of the vigilantes argues that on each previous occasion the Kerinci visited their village a tiger killed someone straight after he left.[28]

Charles Shuttleworth, who led police patrols in the peninsula during the 1948–1960 Emergency, then ran safaris for scientists, photographers and hunters, did state in 1965 that there were cases on record of Kerincis being killed by Malay villagers following attacks by man-eaters, but unfortunately did not cite any. Perhaps he had one of these tiger-trap tales in mind. Alternatively, he may have been thinking of a couple of cases "recalled," in a 1935 article in *Asia*, a popular American magazine, by "Patrick Alexander," which the magazine said was the pen name of "a former officer of the British Criminal Intelligence Department in Singapore" who spent more than thirteen years in the East. In the first of these, "Alexander" claimed that he once attended the trial in Kuala Lumpur of a Malay man accused of murdering a Kerinci pedlar. The Malay freely admitted killing the man, but denied it was murder. The Kerinci, he explained, had sought shelter in his house during a storm, but despite having trekked all day through the jungle, the pedlar ate almost none of the fish and rice he offered him, "and acted sleepily as one who is filled with meat." When the Malay then started telling him about his son, who had gone missing in the jungle only the day before, the Kerinci vomited up a charm the lad had been wearing around his neck, plus a piece of mother-of-pearl he had been carrying, and the pedlar's eyes started gleaming. Said the Malay to the judge: "I then knew that he was a man-tiger and about to turn himself into the animal body again so I took my parang and killed him. What would the Tuan do?" What the judge did, claimed "Alexander," was to sentence him to only two years in prison.[29] Perhaps that was the going rate for the life of a Kerinci pedlar.

Or perhaps "Alexander" simply made the whole thing up, for, mimicking Maxwell, he went on to write that, one day, on the way back from a hunting trip in the mountains on the border of Pahang and Terengganu, he and two friends, one a man called King, just happened to stop at a village and be greeted by the datok, who, along with eleven other men, King just happened to have arrested some years previously on a charge of murdering a Kerinci they had caught in a tiger trap, and so on. As "Alexander" tells it, the datok's name was Kassim bin Abdul Rias. A man-eater was plaguing the area. They set a trap baited with a goat. They caught a Kerinci. The goat was partially eaten. The Kerinci pleaded his innocence. He had only eaten the goat because he had been trapped for three days and had grown hungry. They found paw prints all around. They killed the Kerinci. King came and arrested them. And then, claimed "Alexander," they were all twelve acquitted by the court at Kuala Terengganu.[30]

Returning to reality, an interesting detail of Clifford's brief version of the tiger-trap tale is that Haji Abdallah was found naked in the trap. Clifford also referred to Kerincis who "have left their garments and trading packs in thickets, whence presently a tiger has emerged," which also suggests they disrobed before transforming.[31]

Whether Haji Ali did likewise, however, is not clear from Clifford's text. Neither Maxwell nor Lumsden Milne mentioned their victims' attire, but the assumption is that they were fully clothed. This is what you would expect of an innocent man seeking refuge in a trap, but it would not necessarily have indicated innocence to a Malay.

One night, wrote Maxwell in a story he claimed he heard near the source of the Selim River, someone knocked on an isolated Malay family's door. When the man of the house asked who was there, a voice replied that they were a party on the way to visit friends, but that their torches had gone out. While the suspicious Malay interrogated the stranger through the door, his two boys slipped out the back and sneaked along the ground between the stilts of the house to check out the visitor. There, on the ladder, stood a man talking to their father, "but even while he spoke," wrote Maxwell, "a tail striped in black and yellow dropped down behind his legs, and then up and down his lower limbs ran successive ripples of change and colour. The toes became talons, the feet turned to paws, and the knee-joints, already striped with the awful black and yellow, were turning from front to back." The enterprising lads grabbed the dangling tail and yelled the alarm, but before their father could reach for his spear the creature, now nearly all tiger, tore itself free and fled into the jungle.[32]

Shuttleworth recorded his own version of this tale. Like Locke, he noted that belief in weretigers persisted in parts of the peninsula even in his day, particularly on the remote northeast coast in Kelantan and Terengganu. Malays said, for example, that when Kerincis visited Malay kampongs to peddle their wares they sat outside near a hole in the ground, where they could hide a tail should one start to grow involuntarily. Shuttleworth heard of two such pedlars who once squatted outside a house and started haggling with the lady occupier. The woman's young son came out to answer a call of nature, and on his return noticed a tail protruding from the sarong of one of the pedlars and dropping into a hole. Thinking quickly, he ran up and tied it to a stilt. "A great cry ensued," wrote Shuttleworth, "and before the eyes of the horrified villagers the merchant, with great spasms and contortions of his body accompanied by loud roaring and snarling, commenced to change into a tiger." It must have been a slow process, for the villagers had time to fetch weapons and kill him before it was complete.[33]

Shuttleworth's tale is notable in saying Kerincis were involuntary weretigers. From Swettenham in 1895, through Clifford in 1897 and Maxwell in 1907 (though like many others—surveyor Ambrose Rathborne, for one—Maxwell merely parrots Swettenham) to Bradley in 1929, all portray Kerincis as willing weretigers. (J.B.H. Thurston is more ambiguous, writing only that the mob declared the Kerinci in their trap to be "accursed of Allah.") Yet as early as 1839, English army officer Thomas Newbold wrote of peninsular Malays: "They will point out men that have the faculty of transforming themselves at pleasure into tigers, *or are doomed nightly to become tigers* [author's emphasis], returning to their natural forms by day.... These men-tigers, I must add, were always absent when I expressed a wish to witness the performance of the metamorphosis." Wessing states categorically that he thinks the Kerinci were supposed to be hereditary, involuntary tigermen.[34]

Another report supporting this idea came from Wavell, who related how Barbara Copeland, a Red Cross nurse who lived among Kerincis settled at Ketari—not far from Bentong in Pahang—told him she had heard all the local gossip about Kerincis being weretigers, but was not at all worried for her own safety: even though she heard tigers every night, and often found their paw prints at the end of her yard. When Wavell mentioned this to Malay work colleagues, one of them told him the following story about a friend of a neighbor who married a Kerinci cloth-pedlar as a young woman.

The couple lived on the edge of the jungle near Kuala Kangsar in Perak. Every night, after his evening meal, the man went out into the dark. When his wife eventually confronted him about this, he made various excuses before finally telling her the truth. Every evening, he said, he felt a longing to go into the jungle: "I can't help it. There's nothing I can do. And then ... and then I just change into a tiger!" Understandably, his wife did not believe him, and wanted to accompany him into the jungle that night. "No," he said. "When I'm a tiger I'm not myself. I just attack anything I see and eat it up. I'd eat you up too without knowing that you were my wife." But she was insistent, and reluctantly he agreed, provided she tied herself to a tree high out of reach. This done, he began to turn into a tiger (starting with his head, incidentally, followed by his shoulders, body and legs). Then something went wrong, for squirm as he might his tail just would not grow. At this his wife laughed so much that her ropes worked loose, and she fell to the ground; whereupon her husband-tiger tore her to pieces and devoured her. When he resumed his human form the next morning and found his wife missing, all he could remember was taking her to the jungle, but he realized straight away what he must have done. "He was deeply upset," said Wavell's colleague, "and soon afterwards moved out of the district." Well, he would have had some explaining to do had he hung around.

Anyway, on hearing this so soon after talking to Copeland, Wavell (Figure 11) decided that the Kerincis at Ketari would make a fascinating subject for a weekend's investigation, so off he went one Saturday afternoon with a friend called Gerald Hawkins. So as not to cause any offense, he explained, they decided that the best approach would be a "psychological" one.

On arrival the intrepid pair sought out the headman of the Kerinci community. "He was a tough, stockily built man of unusual strength," wrote Wavell, "who seemed so suspicious of our sudden arrival that momentarily we felt like a couple of intruding game wardens. I, for one, fully expected him to turn into a tiger there and then and rend us both with gnashing fangs." But Hawkins' charm, humility, and knowledge of

6. Tigermen in Malaya

Figure 11. Stewart Brooke-Wavell, known to his friends as "Jungle Jim," and seen here in the field with his recording equipment, wrote that when he hid one night at the edge of a Kerinci settlement at Ketari in Pahang in the Malay Peninsula, he heard no fewer than six different tigers roar in the space of a nervous hour's listening, but sadly they were all just that bit too far away to record on tape (courtesy of Derek Brooke-Wavell).

Malay soon won the man over, and before long the investigators were enjoying his full hospitality.

The Kerincis said they used to peddle cloth, but then had decided to settle down and live like Malays. For a while they settled in Kuala Lumpur, on jungle-clad Bukit Nanas — Pineapple Hill, a small forest reserve — before moving to Ketari. "Over the next few hours," reported Wavell, "the conversation took a series of tactical turns...." But the upshot was that by the end of the evening, when it came time for Wavell and Hawkins to retire to their guesthouse, so subtly "psychological" had their investigation been that they had not even raised the subject of tigers, let alone the delicate matter of weretigers. Still, they and their hosts talked about almost everything else, and a good time was had by all.

Unable to sleep, Wavell got dressed again, slipped outside, and made his way back to the Kerinci settlement, where he hid at the edge of the jungle near Copeland's empty house, the nurse being on leave in England. (A bright moon was shining, he noted.) For a while all was quiet. Then, loud and clear from just beyond the settlement came the *aum*, or roar, of a tiger, followed by that of another, then another. "The next day we returned to Kuala Lumpur," he concluded his tale, "the mystery of the Kerincis unsolved."[35]

One final Kerinci weretiger story — one that has the dastardly foreigner stealing a peninsular Malay man's wife — was that told by Horace "Bill" Harrison in an account of his post–World War I experiences as a tin-mining engineer with the firm of Osborne & Chappel near the town of Gopeng in the Kinta valley in Perak. On arriving in Malaya in 1919 in his early twenties — after three years in the trenches — Harrison took daily Malay lessons after work from the leader of the open-pit mine's tin-washers, an old woman called Hajah Limah, in the course of which she told him the following tale.

A Malay called Jahudin was employed by a neighboring mine to find and fix the leaks that regularly sprung from the miles of pipeline bringing water from a reservoir in the hills through the jungle to the mine. He was an ugly, morose, unpopular fellow, yet had a pretty young wife, Timah, who kept the house clean and cooked an excellent curry. With good reason, then, he guarded her jealously, always making her walk in front of him so he could keep a wary eye on her and any admirers.

One day, wrote Harrison in a passage borrowed almost word for word from Clifford, a Kerinci pedlar called Haji Eusuf arrived laden with sarongs and, taking a fancy to the area, bought a plot of land and settled down to plant coconuts and rubber. Like Clifford's Haji Ali, he was "a quiet well-behaved man, regular in attendance at the mosque for Friday prayers and, as he prospered with his trading, was held in high esteem by his poorer neighbours."

While out hawking door-to-door one day, Haji Eusuf found Timah pounding rice for her husband's supper. As Harrison put it, in his own words this time: "She made an attractive picture as she knelt before the wooden mortar, letting the pestle rise and fall rhythmically; her upstanding young breasts tapered to pointed nipples, rising and falling with the motion of her arms."

Timah loved new clothes as much as any woman, but Jahudin's meager wages left little cash for luxuries. Yet when he came home that evening he found her decked out in a brand-new sarong. "One sarong could be explained," wrote Harrison, "but one alone was not enough to satisfy Timah, and people soon began to talk."

The cuckolded Malay plotted his revenge. The pipeline maintenance gang's families lived in a camp in the hills. The easiest way there was to walk up the pipeline itself; so long as you kept your footing on a tree-shaded section slippery with moss. Jahudin dropped by for coffee one day and happened to mention that Haji Eusuf had some excellent cloth. The families, as Jahudin well knew, had not seen a trader for some time, and they took the bait, saying they would love to see the pedlar's wares. Back home Jahudin casually mentioned this to Timah, adding that it would best if Haji Eusuf went at the end of the day, when everyone was home. The following day, after work, Jahudin came home to change then went out again. And the day after that, the Kerinci was found sprawled on the rocks below the slippery section of pipeline, his skull caved in. He could have fallen, but the police questioned Jahudin. He said he had been with his brother, who backed him up; so that was that.

Life for Jahudin and Timah returned to normal. Then one day when Timah went to the edge of the jungle as usual to gather firewood, a tiger killed her. On hearing the news her grief-crazed husband rushed home, grabbed his kris, and hurried to the scene. Camouflaging himself with palm fronds he lay on his back, pulled her corpse on top of him, and waited, clutching the curved dagger in his hand. When darkness fell and the tiger returned to dine on its victim, he thrust the blade straight up through its belly and into its heart.

Later that night the local English police inspector was reclining in his sarong on a long rattan chair on the veranda of his bungalow, savoring a large whiskey and soda on the rocks, when Jahudin pitched up, covered in blood. Surrendering his weapon, he blurted out, "Tuan, I have killed Haji Eusuf—*again.*"[36]

The story of a man hiding under his wife's body to avenge her killer is an old one in the Malay Peninsula. Skeat heard that the schoolmaster who banished the first tiger condemned it to "catch only the headless." Ever since, therefore, the tiger has had to divine its prey, human or animal, by lying down and gazing at the leaves between its paws: when one of the leaves assumes the headless shape of an intended victim, it knows Allah has granted it their life. At Labu in Pahang, Skeat asked how people knew this, and was told that one day, after toiling all morning in a paddy field, a newly married man took his lunch break in the cool shade of the forest, where he spied a tiger gazing at the leaves between its paws. Creeping closer, he saw one assume the headless shape of his young wife, so he ran back and persuaded the other villagers to escort them both straight home. On the way, the tiger sprang out and killed her. Distraught, he asked to be left alone with her, then lay down, pulled her on top of him, and waited, a kris in each hand. At sunset the tiger returned to claim its meal, whereupon he thrust the daggers through its sides until their points met deep in its heart.[37]

7

Beast People: Weretigers and Wereleopards in India

> *"Those who have seen the tiger when stripped of his skin, can hardly fail to have been struck with the grotesque resemblance to a gigantic human form which is presented by his sinewy and muscular frame as the arms are stretched out on either side. The vast shoulder, arm, forearm, wrist, and hand have a wonderfully anthropoid appearance."* — Joseph Fayrer, *The Royal Tiger of Bengal*, 1875, 8

One time in the early 1900s, when he was stationed in Bilaspur, Best's annual dry-season tour took him back to his old Baiga friend Amoli's home area, around the Chapparwa rest-house in the middle of the Lurmi Range: hills, wrote the Englishman, that were then "all forest, in every direction for mile after mile, forest and nothing but forest of the most delightful kind." On arriving he was disappointed Amoli was not there to greet him as usual. Amoli's fellow villagers explained that the shikari was ill, and took Best to see him in his hut.

Finding Amoli lying coughing and feverish on his bed, Best suspected pneumonia or tuberculosis, both prevalent among the Baigas at that time. Amoli disagreed. Someone had cast a *jadu*, or spell, over him, he insisted. His wife concurred; and said she knew who. It was Lamu, a notorious dabbler in the dark arts who had been thrown out of the village many years before and had lived on his own in an isolated hut a couple of miles off ever since. He was a handsome fellow, she added, but even so, no woman dared go anywhere near him.

Best thought quickly. He pretended to examine Amoli, then pronounced he believed the shikari and his wife were right after all, but luckily he had a charm that could easily take care of the matter. "First Amoli must take this castor oil to cleanse himself of the spirit," he instructed the Baiga's wife, holding up a bottle, "then, if my charm is to work properly," he continued, producing some aspirin, "let him take one of these pills on the second day...." Lastly, announced Best, pulling out some quinine, "when the sweat comes let him take two of these pills and two more each day till all are finished." He would come back after ten days, and in the meantime no one except Mrs. Amoli was to enter the hut. Amoli was to stay put and eat nothing but milk from the sacred cow, and she was to put leaves from the sacred peepul tree on his brow. Finally, he said, "let not the name of Lamu or any other sorcerer save mine be mentioned in his presence. If when he spits he does so into a sal leaf which is instantly burnt, my charm will have still more power." At that Best took his leave.

By the time Best returned ten days later Amoli had made a full recovery. He assumed his usual role as Best's personal shikari, but for week after week Best had what he could only describe as the most rotten luck. Almost everything he fired at he missed, and most unusually he did not bag a single tiger. He put it down to malaria and overwork, though once, when he missed some birds, he joked that someone must have put a jadu on *him*. Early one morning, finally, when he had a clear shot at a stag and apparently missed it completely, he turned to Amoli in disgust and exasperation and asked him what *he* thought was going on. Everyone knew, came the immediate reply. The sahib had been right when he said someone must have put a jadu on him. It had happened the previous year when he killed a stag near the Gond village of Deosara and divided the precious meat among its people. An old woman had complained she had not received her share, and had cursed the sahib out loud. Said the shikari: "We know it is so and it is known to us that the Gond women cast the spell. Let a Baiga of the Bhumias deal with the witch."

Appalled by this idea, Best returned to camp for an urgent meeting with Amoli and the other Baigas, the upshot of which was that they agreed a less drastic plan of action. Cockerels were duly sacrificed, and after breakfast, to test the result, Best had the men beat a nearby nullah, three previous beats of which had failed to yield a particular tiger. Sure enough, out came the animal, which Best dropped with one well-placed shot. The jadu, it seemed, had been broken. That evening he and Amoli even found the carcass of the stag they thought he had missed in the morning.[1]

Colonial memoirs from all over India are full of such anecdotes. In 1878, writing as "Maori," Scottish indigo-planter James Inglis recalled a villager in northern India once complaining to him that his wife had come down with a fever after the woman next door, a well-known witch, had splashed water on her following a quarrel. The man wanted the woman beaten and expelled from the village, but instead Inglis gave him some quinine for his wife, then had his moonshee write some Persian characters on a scrap of paper, which he gave to the man while muttering a rhyme, with the instruction to take three hairs from his wife's head, and some clippings from her toe and fingernails, and burn them outside at moonrise. The woman recovered, and Inglis acquired quite a reputation as a "witch-doctor."[2]

Inglis typically lacked the sensitivity of later good eggs like Best, however, for he also recalled once ordering some villagers to dig over a burning-ghat so he could grow oats on it. When they refused, for fear its many ghosts would haunt them in revenge, he told them he would settle the spirits with a powerful spell. To this end he took a branch from a sacred bael, or wood-apple, tree, as used in Hindu funeral pyres, dipped it in the river, and waved it around his head while walking backwards and chanting "the first gibberish" that came into his head, a nursery rhyme from his schooldays:

> Eenerty, feenerty, fickerty, feg,
> Ell, dell, domun's egg;
> Irky, birky, story, rock,
> An, tan, toose, Jock;
> Black fish! white troot!
> "Gibbie Gaw, ye're oot."

When he then declared his charm had worked, the men willingly set to work.[3]

Such anecdotes were intended to show the benefit of enlightened British rule, as well as to amuse, but the persecution of supposed black magicians was evidently commonplace in colonial India, especially among tribespeople. Even after Independence, Fuchs reported it so among all castes and tribes in eastern Mandla in the late 1940s and early 1950s—particularly among the Gonds and Baigas—while Juliusson reported instances of it among Gonds as recently as the late 1960s.[4]

Just to point the finger was usually enough. Almost every tribal village had its resident soothsayer or witch-finder whose job it was to "divine," by means of a suitably elaborate ritual, the identity of the culprit or culprits whenever some ill fortune not attributed to a deity or spirit was encountered: in other words, whenever "proof" was needed to condemn someone already targeted. According to geological surveyor Valentine Ball, writing in 1880, often the accused readily agreed with their findings, however dire the consequences, presumably on the basis they saw them more as a diagnosis.[5]

As ever worldwide, women, particularly childless old women, were favorite targets. If she protested her innocence she might simply be beaten until she confessed, reported John Campbell Oman, one-time professor of natural sciences at the Government College in Lahore, in 1908. But sometimes she was "tested," and naturally the various ordeals were designed to guarantee a guilty verdict. As in medieval Europe, a common test was to "swim" her, wrote Crooke: if she floated she was guilty, while if she sank and drowned she was innocent. Other tests involved ordeals by fire, such as making her plunge her hands into boiling oil or ghee: if they burnt, she was guilty. In Mandla, Fuchs heard that the hands were not burnt if the oil or ghee really was boiling hot. He even heard of a Gond farmer whose cow and calf had both died within a suspiciously short space of time rounding up all the women of his village and forcing them in turn to "milk" a red-hot crowbar with ghee-covered hands: reportedly none came to any harm and all were declared innocent.[6]

Perhaps they passed such tests in trance rituals associated with worship of their Devi, their supreme Mother Goddess or Earth Mother. Ethnologist Sudha Chandola has witnessed people taking part in such rituals in northern India apparently licking red-hot metal without coming to any harm.[7]

More likely, though, a targeted individual waiting their turn is deceived into believing a test is real, when it is not, and so is prompted to "confess" at the last moment.

At one time the punishment for witchcraft in India was often death, since anything less might allow the culprit to avenge themselves on their accusers. And if they did have any family, noted Ball, they too often suffered the same fate, as witches and sorcerers were widely said to pass their powers on to their offspring.[8]

As with other traditional Indian practices they found abhorrent, like suttee, or widow-burning, in India the British tried to stop the persecution of supposed witches and sorcerers. As reported in *The Asiatic Annual Register* for 1801, the clamp-down followed a case in Patna, the capital of Bihar, on the Ganges, in 1792, when five women there were summarily convicted of witchcraft by a local "tribunal," and put to death. The ringleaders were rounded up and tried for murder before the local circuit court, where they were convicted but afterwards pardoned, several witnesses having testified that such "tribunals" were an accepted practice in India. The government then proclaimed that "all persons who should put any persons to death on the ground of being convicted of sorcery, should be deemed guilty of murder, and the persons forming the tribunal accomplices."[9]

The lynching of supposed witches and sorcerers still took place after 1792, however; especially among tribespeople, and particularly in remote areas like Chota Nagpur. "When during the mutinies," wrote Colonel Edward Dalton in 1872, "the Singbhum District [the Singhbhum district, now in Bihar] was left for a short time without officers, a terrible raid was made against all, who for years had been suspected of dealings with the evil one, and the most atrocious murders were committed. Young men were told off for their duty by the elders; neither sex nor age were spared."[10]

Many tribespeople, it seems, took no heed of the consequences of their actions. Ball recalled an instance from Chota Nagpur. The district deputy commissioner who tried the case told him that on losing some cattle to disease a Kol man hired the local witch-finder to identify the "real" culprit, and a certain old woman was duly pointed out. She readily admitted the offense, but the man let her off with a warning. A while later his eldest son died. Again she freely admitted her guilt, adding that this time she had been helped by three "sisters," who in turn readily confessed. One by one the man hacked off their heads with a sword, then calmly handed himself in, though it meant a certain death sentence.[11]

Oman speculated that many thousands of women must have suffered a similar fate in remote areas, with the local police turning a blind eye. At the same time he had no doubt the Patna proclamation saved the lives of a great many women. And based on anecdotal evidence, it does seem that for fear of the law people more often banished their "witches" and "sorcerers" under British rule.[12]

Witches in India supposedly acquired their powers from the usual suspects. Fuchs heard they learned the necessary spells from their tutelary goddess the Earth Mother, both in their dreams and, in the fortnight following Diwali, while wandering around naked all night in a trance under the cover of darkness provided by the new moon.[13]

Diwali and Nauratri, the nine nights beginning March 29 devoted to the worship of Durga, were when witches generally were said to be most powerful. Or, as Major-General Sir John Malcolm put it in 1823: "When the fit is on them, they are sometimes seen with their eyes glaring red, their hair dishevelled and bristled, while their head is often tossed around in a strange convulsive manner. On the nights of these days they are believed to go abroad, and, after casting off their garments, to ride upon tigers and other wild animals; and if they desire to go upon the water, the alligators come, like the beasts of the forest, at their call, and they disport in rivers and lakes upon their backs till near dawn of day, about which period they always return home, and assume their usual forms and occupations."[14]

Magic powers have to be paid for, and Fuchs heard that at night, when ordinary law-abiding folk were asleep, such a woman prostrated herself naked before an image of the Earth Mother and promised offerings of human blood. With the help of a spell she then turned herself into a cat, rat, or worm, in which guise she sneaked into the home of an enemy — invariably a man — and sucked his blood. She visited the same victim every night for week after week until it killed him. But his problems did not end there. If he was cremated, his body would not burn. Cremated or buried, he remained at the mercy of the witch, who now fulfilled her promise in a particularly ghoulish manner.

Naked again, a lamp in each hand, and on her head a brass plate with another lamp and small pots of food, water, and oil on it, in the dead of night she walked around the grave or pyre three times, anointed herself with oil, then dug the corpse out of the ground or ashes, washed it from head to toe, and revived it with powerful spells. When, zombie-like, it sat up, she plied it with food and drink and began beating and interrogating it. The moment it groaned or cried out, it was fully in her power and ready for the final sacrifice. Hacking off its head she drank deep of its remaining blood before proceeding, still naked, to the shrine of the Earth Mother. There she vomited up the blood over the image and performed a little dance for her deity. Finally she offered the goddess the blood of a cockerel or crab, plus a few drops of her own blood, and beseeched her to grant another victim soon.[15]

Never mind cats, rats, and worms: Indian witches and sorcerers alike were widely said to be particularly fond of turning into tigers and leopards for their own nefarious ends. Take the following case reported by Dalton, one often cited by popular authors of the occult. A Kol man tried for the murder of another Kol, a man called Pusa — a deed the accused readily owned up to — stated in his defense that on the day in question a tiger killed his wife in front of his very eyes, after which "he stealthily followed the animal as it glided away after gratifying its appetite," and saw it enter Pusa's hut. Summoning Pusa's relatives, he told them what had happened, and declared that Pusa and the tiger must be one and the same. To his surprise they agreed, saying they had long suspected such. (One night, they attested at the trial, Pusa roared like a tiger while devouring a whole goat, while on another occasion he told his friends he longed to eat a certain bullock, and that very night a tiger took the animal.) They went inside Pusa's hut, where sure enough "they found him and not the tiger," tied him up, and handed him over to the defendant for immediate dispatch.[16]

As for witches, in the Godavari district in the south of the country, reported Edgar Thurston, they were known as *chedipes*, or prostitutes. A chedipe spent her nights riding a tiger around the neighborhood, when not doing the usual thing of sucking a man's blood as he slept (by the unusual means of inserting a big toe in his mouth). And sometimes she undressed behind a bush where, assuming the form of a *marulupuli*, or "enchanting tiger"—a tiger with one, presumably shapely, human leg—she lay in wait to lure men to their doom.[17]

As is the way, sometimes there was a frenzy of finger-pointing. In 1885, Sir James MacNabb Campbell reported that when a certain John Campbell became one of the first British officials to visit the small princely state of Surgana in the Western Ghats north of Deolali in 1871, he found a widespread "fear" of witches there. Whenever a man was taken sick his family hid him away, in case one of the many local witches heard about it and decided to come after his liver. The "fear" even spread to several of Campbell's servants when *they* fell ill. Then one day Campbell was shown a report written by a local Brahman official, and "gravely" filed alongside all the other state reports. According to this, for a long time past tigers had been taking a heavy toll of the men of the area with no one able to bring any of them to book. As a result it was widely said the killers were witches in the form of tigers, one man telling Campbell: "They could not have been real tigers, because no one could find out anything about them. They had no regular beat, and no haunts or lairs. They appeared, killed a man, and disappeared in a way real tigers could not have done...." Eventually the Brahman himself became convinced, and in each village where a man had been killed he issued a proclamation that if any more such deaths occurred he would arrest all the women of the village and send them to nearby Dhule to be dealt with by the British government. He noted that the proclamations had worked with immediate effect, with no deaths recorded in the weeks since: and concluded by saying he trusted his actions would meet with the approval of his superiors.[18]

More typical, perhaps, was the case of a tiger Hanley encountered as a young man at the outset of his career as a tea-planter—after several years in the Indian Army—at Naganijan in Assam. This time the purported motive was not a craving for human flesh and blood—the tiger in question was not even a man-eater—but merely murderous revenge.

One Sunday morning soon after his arrival, Hanley rashly decided to collect some rare "black" orchids he had spotted growing at the edge of the forest in the Pootinadi section of the plantation. Bending to his task he heard a rustle in the tall reeds and elephant grass; and looked up to see a tiger staring straight at him from only a few feet away. The tiger bared its teeth and snarled, but Hanley had the wits and courage not to turn and run but to try to stare it down. At this he succeeded and the tiger eventually retreated into the depths of the forest.

When Hanley subsequently arrived late for lunch, his worried plantation manager, an old hand called Mr. Percy, chided him for his foolishness, saying even the jungle-wise immigrant Santhal and Oraon plantation coolies would not dream of venturing into the Pootinadi section alone and unarmed, for it was the known haunt of Bengala, famed throughout the district as "the wise and cunning one." Scores of shikaris and sportsmen alike had been trying to shoot it for more than a decade. Hanley had had a lucky escape, for while Bengala was a cattle-killer and not a man-eater it would not have hesitated to kill him if it had felt cornered, having once killed a beater in order to break through the line.

Unsurprisingly, that Bengala was able time and again to defeat the best efforts of hunters had earned it a supernatural reputation, one Oraon coolie later assuring Hanley that Bengala was "not an ordinary tiger, but a ghost tiger, a devil in the form of a tiger, and the bullet is not made which will kill him. And his size, *huzoor*! Wah! He is the largest tiger in Hindustan no doubt. Truly a king among tigers." Hanley subsequently heard numerous stories about Bengala's prowess, how the tiger never failed to make its kill. At first he dismissed these out of hand, knowing people everywhere were prone to exaggerate the attributes of any big cat in their vicinity. But after watching Bengala stalk cattle on a number of bright moonlit nights — nights when the full moon made all "as bright as day" and lit up the tiger's eyes like lamps — he changed his mind, so impressed was he by its sheer size and beauty, its great stealth and awesome power.

Then, late one Sunday night — again, a night of the full moon — two Santhal coolies, Mujo and Rama, were staggering home from a pay-day party at which the rice beer had flowed freely as usual when they spied Bengala crossing the bridge over the Pootinadi just ahead of them. Mujo, who was much the worse for drink, let loose an arrow, and more by luck than judgment the missile found its mark, striking the tiger in the back. At this Rama sensibly ran and hid behind a tree, but Mujo stood his ground and was still trying to fix another arrow in his bow when the enraged animal sprang and sent him flying through the air with one swipe of a paw. The blow snapped Mujo's neck and tore half his face away. Sniffing the body, the tiger shook the arrow from its back and padded away into the night.

Rama woke Hanley in his bungalow, and gun and flashlight in hand the young planter followed him to the bridge. He saw straight away that Mujo was dead, but ordered the coolies to take him to the plantation hospital anyway. Rama then surprised Hanley by telling him the Santhal woman Teara, a sinister crone who everyone on the plantation, coolies and Indian servants alike, knew was a witch, had actually killed Mujo, having "changed her form" for the task. Hanley told him he was talking drunken nonsense, to go home and get some sleep.

Hanley himself had just nodded off again when he heard a voice urging him, "Awake, *huzoor*, awake! I am in much danger." It was Teara, standing right by his bed. Hanley shouted angrily through his mosquito-net for his chowkidar, but as usual the man was sound asleep. Before he could bundle her out, Teara hastily explained that the other Santhals had torched her hut and were now after her blood; adding that the jungle spirits she "controlled" had told her only he could save her. Hanley looked out the window. Sure enough her home was ablaze.

Ordering his servants to protect Teara, Hanley hurried over to find a crowd of Santhals standing around watching her hut burn. Naturally none of them knew how the fire had started, but the senior headman told Hanley they were all "convinced" Teara had killed Mujo. A week earlier, when drunk as usual, Mujo had threatened to burn her alive for being a witch, at which she had warned him it was he, not she, who would die, saying, "I will be one with the jungle, and from the jungle death will come to thee."

At this Hanley issued his own warning, telling the headman that if anyone killed Teara he would personally see to it they hanged. Someone would kill her all the same, replied the man, and in such a way that left no evidence: the only answer was for the huzoor to send her home to the Santhal parganas.

Realizing he was right, in the morning Hanley gave Teara 50 rupees and put her on the first train home. (To recoup his loss he later fined the other Santhals for torching her hut. They paid up without complaint.) As she boarded the train she told him her jungle spirits would protect him from that day forth. "The train steamed out and she was gone," he wrote, "but like the Santhals, I began to believe that she really was a witch."[19]

Whether the Santhals actually believed Teara to be a witch is a moot point. Most likely they just took advantage of Mujo's death to get shot of her.

In some ways Teara was lucky. According to Crooke, convicted "witches" and "sorcerers" in India sometimes had their noses cut off, or their front teeth knocked out. The latter was usually said to be to prevent them uttering spells, but Crooke reckoned it was to render them harmless in case they tried to avenge themselves in tiger or leopard form: a sort of pre-emptive, reverse wound-doubling.[20]

It is not clear if Teara's supposed shapeshift was physical or spiritual, but in another case, reported by diamond and gold prospector and mining engineer A. Mervyn Smith in 1904, the supposed shapeshift was clearly the latter, with robbery, revenge, and a lust for human flesh and blood all alleged motives.

When Mervyn Smith moved into a bungalow in the village of Somij in Chota Nagpur sometime in the second half of the nineteenth century he learned that a man-eating leopard had been terrorizing the neighborhood for three years past; and shortly afterwards he befriended a baghmari from the nearby village of Bara, a man called Beema, who told him the full story as he saw it. The true culprit, said Beema, was an old woman from his village called Lagon, whom everyone "knew" was a witch who could turn herself into any animal she liked to harm her enemies. And if the sahib did not believe him, he could ask the village headmen.

Three years before, explained Beema, he had been but a lowly weaver. Then one day Lagon asked him to make her some cloth, which he did, using one rupee and two annas worth of cotton. But when he quite reasonably tried to charge her six annas extra for it, she refused, and when he protested at this tightfistedness she cursed him, saying a "tiger" would eat him: "The curse was a great curse and made with bent fingers, and her great toe marked the curse on the sand. After this I was afraid to go to the jungle alone, as I was always in dread of tigers. I killed a cock and sprinkled the blood round my hut, yet the witch's curse was strong and I felt the water on my back...." Then a few nights later a leopard slipped into Beema's goat-pen, slaughtered two goats, and was just carrying off a third when Beema burst in and swung his ax, severing part of one ear and striking its left forepaw. In return the leopard clawed his scalp, a wound that later caused most of his hair to fall out.

Beema said he "knew" full well Lagon had "entered" the leopard to do him harm. So when she called by his home the next morning, told him she had heard about the attack, and offered him four rupees for the piece of ear, saying she wanted it to make medicine, he refused. At this she closed one eye and again marked the sand with her big toe. "Then I knew it was a question of her life or mine," said Beema. He took the piece of ear straight to a Gond man called Gagee and paid him two rupees to turn it into a lucky charm. Thereafter, Gagee assured him, as long as he had the charm on him at all times Lagon could not hurt him. Now confident of his safety, Beema vowed revenge on the woman, to which end he became a bagh-mari that very day.

For some time after that, said Beema, Lagon was "ill" and stayed in her hut: in other words, recuperating from the repercussion wounds he had inflicted on her while she was "inside" the leopard. Then one morning the women of Bara went out to harvest rice and one of them left her baby in the shade of a tree. When they broke for a rest at noon the infant was nowhere to be seen. Drops of blood led to the paw prints of a leopard; minus those of the left forepaw.

Thereafter, said Beema, barely a month went by without this three-legged "devil" carrying off children from Bara and the neighboring villages of Dalki, Derwa, and Huthutwa, not to mention Somij. Young goatherds were particularly at risk when they grazed their animals in the forest. Soon the man-eater grew so bold as to kill women too, and before long not even the men dared to leave their huts after dark.

One moonlit night the leopard sneaked into Bara and carried off a 14-year-old girl. The next day villagers found her feet and part of her chest and head near Lagon's hut. "The old witch was examined," said Beema, "and it was found that she, who had previously been all bones, was now sleek and fat." An earlier victim's jewelry was also discovered in her hut. Lagon claimed she had found it while out gathering firewood, but the villagers dismissed this explanation out of hand, and only fear of the law prevented them from burning her alive on the spot. Instead they burnt down her hut and banished her, the village chowkidar escorting her to a relative's home in a distant village.

But the killings continued. Experienced shikaris were called in, yet the man-eater managed to avoid all their cunningly laid traps. Sportsmen were asked to help, but though they shot several tigers and leopards in organized beats the killer remained at large. The government offered 50 rupees for its destruction, but the reward went unclaimed. "When we looked for it in Bara," said Beema, "it was heard of in Derwa, and when we got there, it was back again at Bara." Villagers sacrificed cockerels, goats, and even a buffalo calf, all to no avail. Fearing for their lives, many families abandoned their homes and fled.

Beema by then had become an accomplished bagh-mari, setting ingenious crossbow-traps not only for the man-eater but for any livestock-lifting tigers and leopards in the area. Before setting a trap he prepared two poison-tipped arrows. Usually he bought an aconite root from a medicine store in Chaibasa, ground it up with some boiled rice, and rubbed the paste into rags, which he wound around the arrowheads just behind the barbs. When possible, though, he preferred to soak his rags in cobra venom, which unlike aconite, he said, retains its potency indefinitely. This he milked once a month from two captive cobras by provoking them to bite a ripe plantain on the end of a stick.[21]

Beema set a trap by mounting his crossbow, armed with both arrows, on vee-shaped sticks some eighteen inches off the ground on one side of a path used by his quarry, then stretching a string from the trigger to a stick on the other side. When the chest of a tiger or leopard touched the string, the arrows shot into its flank. He also angled two higher strings, so any bullock or person coming along the path from either direction would harmlessly trigger the trap before reaching the lethal lower string. (For the trap to work, said Beema, it was first necessary to sacrifice a white cockerel.)

In this way Beema accounted for several leopards, tigers, cheetahs, and bears—the skins of which Mervyn Smith subsequently bought to hang on his veranda—but not the man-eater: until one night shortly after Mervyn Smith moved into Somij, when it killed a young man in his hut and dragged him off to a path behind Mervyn Smith's bungalow and there ate all of him bar his head and legs. Here Beema set a trap and killed it at last. It was lame in the left forepaw and missing part of one ear. Mervyn Smith duly added its skin to his collection.

For Beema, the death of Lagon a week or so later was conclusive "proof" of her guilt. "May her bones be accursed!" were his final words on the matter.[22]

Mervyn Smith remarked that he could not vouch for the truth of Beema's story about the supernatural origins of the man-eater, though other villagers declared it to be true. Less charitable minds might suggest the weaver-turned-bagh-mari helped create the killer himself—albeit inadvertently—by incapacitating it with his ax: and that the headmen and other influential villagers then took advantage of its presence to get rid of Lagon.

Be that as it may, that Lagon died a week after the leopard was killed confirms her supposed shapeshift was spiritual. The beauty of accusing someone of spiritual shapeshifting is that they cannot defend the charge, for even being seen to be asleep in bed at the time of an offense is no alibi. Yet in every tale I know of from India, when *men* are said to willingly become tigers or leopards, they physically transform. And quite logically the big cats in question are male animals: ones, moreover, that are man-eaters of the very worst kind. Equally, the chedipes of Godavari notwithstanding, I know of no man-eating leopardess or tigress in India, however bad, being said to be someone physically transformed.

The fundamental flaw with saying a man-eater is a certain man physically transformed is that it is so easily disproved: for being seen to be asleep in bed at the time of an offense is an unshakeable alibi. Likewise the whole charge goes up in smoke if the man-eater is shot dead but the "suspect" remains all too alive and well: unless, that is, you conveniently have him disappear at the outset.

Like Dalton's brief but succinct tale of the dastardly Kol Pusa, another Indian weretiger story often cited by popular authors of the occult is a longer, more colorful yarn told by Irish author Elliott O'Donnell (Figure 12) in his book *Werwolves*, first published in 1912.

London-based O'Donnell was obsessed with all things occult, "evidence" for which he discovered on an almost daily basis. Reading his later *Strange Cults and Secret Societies of Modern London*, one gets the impression he was incapable of stepping outside his front door without encountering some previously unreported instance of the bizarre. In

Figure 12. A 1930 photograph of Elliott O'Donnell (1872–1965), the author of a fanciful "true" story about an Indian weretiger, published in *Werwolves* (1912). In *Strange Cults and Secret Societies of Modern London* (1934) he even claimed that African leopardmen and women were at large in the English capital (© National Portrait Gallery, London).

Werwolves, meanwhile, he recounted with relish one gruesome case study after another, many of them remarkable mainly for their misogyny.[23]

O'Donnell's weretiger tale is a supposedly eyewitness account concerning the Khonds, or Kandhs, of Orissa, a tribespeople once notorious for making gory human sacrifices to their Earth Mother, Tari Pennu, a highly vindictive goddess identified, as usual, with Kali or Durga. Accounts vary, but according to those gathered by Edgar Thurston, it seems that every year, until the British stopped them in the early 1800s, the Khonds took a *meriah*—someone raised or captured for the task—drugged them, or broke their bones, to prevent them resisting, dragged them around the villages, where people scrabbled for drops of their saliva, then hacked them to death, strangled them, or roasted them alive, before scattering bits of their flesh over the fields to ensure a bountiful harvest. The more blood the victim shed, the redder would be the turmeric crop; and the more tears they shed, the more rain Tari Pennu would send. If the sacrifice was not made, Tari Pennu would destroy the crops and send disease and man-eaters to ravage the people.[24]

According to Thurston, the Khonds believed that the soul, or some aspect of it, wanders abroad while you sleep—wanderings that manifest themselves as dreams—and that some of their kind possessed the power to send out their souls "changed into tigers." These tigers then sometimes devoured the wandering souls of other sleepers, causing their victims to fall ill, but sometimes, when not scoffing livestock, they physically killed—and presumably also devoured—their sworn enemies.[25]

According to O'Donnell, however, the Khonds claimed that some of their kind could physically change into tigers, and many people who had traveled in Orissa had assured him this was true. The power was hereditary, but could also be acquired, the Khond tiger god, who had the welfare of the Khond people at heart, bestowing it as an honor and a privilege on anyone supplicating it with sufficient devotion and earnestness; in the wilds of the jungle at sunset, generally. One informant, a "Mr. K—," had even seen it happen, apparently.

O'Donnell wrote that this Mr. K— told him he had been staying in a Khond village one time when one evening he hid by a clearing where he had heard such transformations took place, and there awaited the arrival of a would-be weretiger. Sure enough, along came a barefooted young man. Indeed, "He was hardly more than a boy—slim and almost feminine—and came gallivanting along the narrow path through the brushwood, like some careless, high-spirited, brown-skinned hoyden." But the

7. Beast People

instant he reached the clearing his manner changed. Stepping into the moonlight he knelt facing a giant kulpa briksha tree, bowed his forehead to the ground three times, then looked up while chanting some strange repetitive refrain. An unnatural darkness swooped down from the distant hills (the Eastern Ghats, presumably), the jungle fell eerily silent, and Mr. K — was filled with a sudden dread.

Then the silence was broken by a horrible cry, half animal, half human, at first faint and far off, but quickly growing louder and nearer. Mr. K — heard something striding towards the clearing; and in burst a column of crimson light some seven feet tall and one foot wide.

The lad rolled his eyes and gasped. Stuttering with terror he scratched a mark in the ground and laid a bead necklace (representing Shiva, perhaps?) over the mark. A shaft of light shot out from the base of the column onto the beads, which shone bright red. The young man donned the glowing necklace and began vigorously clapping his hands while uttering a succession of increasingly animal-like shrieks that culminated in a terrifying roar. The column vanished, the darkness cleared, and there, in a flood of brilliant moonlight, in place of the boy Mr. K — saw peering up at him the yellow eyes of a bloodthirsty tiger.

For a moment Mr. K — was paralyzed with fear, then the screech of a bird brought him to his senses, and he turned and fled as the weretiger gave chase. Ahead he saw two tall trees: a tamarind and another kulpa briksha. Instinctively he headed for the latter, the weretiger closing in on him all the while; but purring, as if content it could catch him any time it chose. From the patter of its human footsteps he realized its transformation was not yet complete.

With a last lung-bursting spurt he reached the tree and made a desperate lunge for the lowest branch, some eight feet from the ground. Even as he hung by his fingertips a paw shot past his face, and he resigned himself to his fate. But then, extraordinarily, the weretiger growled in terror and bounded away into the night, leaving an unharmed and mightily relieved Mr. K — to make his way back to the village.

The next morning the villagers found the mutilated, half-eaten bodies of an entire family — man, wife, son, and daughter — on the floor of their hut. All had been sworn enemies of the young man in the clearing.

When Mr. K — told a village elder of his escapade, the man said he knew the tree well, and the sahib undoubtedly owed his life to it. He explained about the original kulpa briksha, how the names of Rama and Sita were written on all its descendants on Earth, and how anyone who touched one was safe from any animal. As for the events in the clearing, only those like the boy who had been initiated into the full magic rites in their youth saw the tiger god clearly when it materialized; as a huge creature, half man and half beast. Those like Mr. K — who were to some extent clairvoyant saw it as a column of crimson light, while those who were in no way clairvoyant saw nothing. The boy had prayed to the tiger god to make him a weretiger so he could avenge himself in the cruelest way possible on his enemies. But now he was condemned to transform every night, and would likely continue to kill and eat people until someone was brave enough to shoot him dead. It would be best, then, for Mr. K — to leave immediately.

This Mr. K — did, but on his way to his home in the hills he returned to the tree in the twilight. Softly traced on the silvery bark, as if by some supernatural hand, was the name Rama, in Sanskrit. That was enough for him. Checking his gun was loaded he hurried on; and never went near the place again.[26]

Somewhat inconveniently for anyone else who might have wanted to interview "Mr. K —," O'Donnell remarked that he had subsequently seen his name on a list of missing passengers following the *Titanic* disaster.

Also, O'Donnell's description of the kulpa briksha suggests not a palmyra, but a peepul, a fig tree that indeed has silvery bark, and branches, as opposed to fronds, that someone running for their life might reach from the ground. A peepul usually grows as an "air plant" on another tree, eventually encasing and killing its host as it sends down branches to root in the soil. As its scientific name, *Ficus religiosa*, suggests, it is highly sacred, to Buddhists as well as Hindus. In O'Donnell's defense, it is most sacred to the former when it grows on a palmyra, and just as the gem-laden branches of the kulpa briksha are said to cast a crimson glow on the ground, so the palmyra and peepul both bear bright purple, if not crimson, fruit. Perhaps, then, O'Donnell's kulpa briksha was a peepul encasing a palmyra.

O'Donnell's story may be of doubtful authenticity — to put it mildly — but it does offer indirect support for those who regard yarns about teenage male werewolves as handy metaphors for the changes wrought by puberty.

Back to reality, one of the worst Indian man-eaters on record was a leopard that terrorized a wide area around the village of Kahani, near Dhuma, in the Mahadeo Hills on the edge of the Narmada valley in northern Seoni, from 1857 to 1860. One colonial who witnessed its terrible exploits firsthand as a young district officer was Sterndale, who eventually ended up, for the last few years of his life, as governor of the island of St. Helena in the South Atlantic. He wrote about it in both *Seonee* (1877), a dramatized book for boys in which he cast himself as "Major Fordham," and *Natural History of the Mammalia of India and Ceylon* (1884), which, like many such Victorian titles, was essentially a shooting guide.[27]

Sterndale put the Kahani killer's tally at more than 200 victims, though Forsyth, who passed through the area in the early 1860s, reported a more modest but still impressive total of just under 100.[28]

Like many nineteenth-century sportsmen, incidentally, Sterndale thought there were two different varieties or even species of leopard or panther: a smaller one that ate small prey; and a larger one that ate larger prey and sometimes took to man-eating. Many colonials called the former the leopard ("*Felis leopardus*") and the latter the panther ("*Felis pardus*").[29]

Sterndale, however, considered the name "leopard" confusing, as it was also applied, in the term "hunting leopard," to the cheetah, so he muddied the waters further by distinguishing between the smaller "panther," which he termed "*Felis panthera*," and the larger, more powerful, much more dangerous "pard," "*Felis pardus*." (Naturally he identified the Kahani killer as the latter.)[30] Nowadays only one species is recognized: *Panthera pardus*, the leopard or panther.

Whatever the Kahani man-eater's true toll or identity, for three years it spread fear far and wide. Its many victims included Gonds guarding their precious crops at night from inside their flimsy, tepee-like bamboo shelters. One night it leapt through the fire at the entrance of one such shelter, seized the man sleeping inside by the throat, and dragged him out through the flames. His young wife woke just in time to grab his legs, at which a terrible tug of war ensued in front of their terrified toddler, the woman shrieking all the while, until at last the leopard let go and fled. Sadly, her desperate efforts were all for nought, for her husband was already dead. Sterndale found her weeping over his body the next morning.

Nor were villagers safe in their homes. Sterndale noted that in the village of Sulema alone it was said to have claimed no fewer than forty men, women, and children as they slept uneasily on their beds.

Following an attack in a ravine near the start of the man-eater's career, Sterndale, who was not then twenty and had been in India barely a year, organized a beat without first examining the scene, under the usual assumption the culprit was a tiger. Three times the leopard broke cover — the first time actually pausing in front of Sterndale's elephant — and each time both Sterndale and his brother-in-law, Colonel Thomson, held fire. "I felt very much inclined to knock him over," wrote Sterndale, "but I thought of the chance of losing the tiger, and so let him pass on." Only when they belatedly examined the victim's remains did he realize his mistake, but by then it was too late, and shortly afterwards Sterndale was posted away to Sasaram in Bengal (in what is now Bihar) to help suppress the Mutiny as a member of the Irregular Corps.[31]

When he returned two years later, however, and heard that the man-eater was still at large, he was determined to atone for his error. "I was out several times after this diabolical creature..." he recalled. "All day long I scoured the country with my elephant, all night long I watched and waited." His nervous followers daily imagined seeing the leopard in camp — a sentry swore he saw it glowering at him across the fire one night — but Sterndale never so much as glimpsed the animal, so typically "cunning" had it become. Forsyth wrote that sportsmen would spend all day fruitlessly searching for it, return exhausted to camp at nightfall, then wake the next morning to find its paw prints all around their tents.

Initially it had been incredibly bold, killing even in broad daylight. Sometimes, heard Forsyth, when it was chased away from one end of a village it took advantage of the confusion to sneak round and snatch someone from the other end. But now, having been repeatedly shot at, it was extremely wary — cowardly, Sterndale called it — killing only at night, and often being disturbed by the slightest sound into abandoning a kill. As a result, wrote Sterndale, it sometimes killed two or even three people in the course of a single night, often in places several miles apart. As he put it, it seemed to kill for killing's sake, often leaving the bodies untouched. Forsyth went so far as to state that it *never* actually ate any of its victims, but merely lapped the blood spilling from their throats.

Unsurprisingly, the story spread that the creature was really a wereleopard. According to Sterndale, the culprit was said to be a missing Gond called Chinta who had lived with his wife in an isolated hut in the jungle and had been a powerful sorcerer who could turn both himself and others into wild animals. He had been feared by all, and no one dared refuse his frequent requests for flour, millet, mahua spirit, and even that much prized commodity salt. Rumor even had it that tigers were seen taking offerings of meat to his hut. Noted Sterndale, "certain it was that the flesh-pot often simmered on his fire, when the rest of the Gonds had but grain and roots to live upon."

The story went that one day Chinta was walking through the jungle with his wife when they spotted some nilgai — a large antelope — and she expressed a desire for fresh meat, having eaten none for days. Chinta took a dried root from his pouch, and urging her not to be afraid told her he was going to turn himself into a leopard to kill one of the nilgai, and needed her to hold on to the root at all costs until he approached her in his leopard form, at which she was to hold it out for him to smell to reverse the transformation. Satisfied she was clear on these critical instructions, Chinta disappeared into the bushes. A short while later a leopard sprang out, killed a nilgai, then bounded towards her, its mouth dripping blood. Losing her nerve, she flung away the root and ran for her life. Her leopard-form husband searched and searched, but could not find the root. Enraged, he ran her down in their hut and tore her to pieces on the spot. Thereafter he dedicated himself to avenging all mankind, whose form he could never now regain, as the dreaded man-eater of Kahani. (Disappearing into the bushes before transforming is a common theme in physical-transformation yarns — de Groot's Jiangxi weretiger, and Edgar Thurston's chedipes, both did it, for one — the implication being that the shapeshifter is hiding their clothes.[32])

Presumably either the real man-eater actually killed Chinta and his wife, or — what is more likely — some person or persons unknown took advantage of it to dispose of the Gond (and maybe his wife too). Some people undoubtedly used the cover of man-eaters to commit murder. A cynic would even argue that it was sometimes easier for the police to blame a big cat than to bother looking for the real culprit, especially if the victim was an Untouchable.

Murderers sometimes tried to disguise their handiwork as that of a man-eater whether a real one was on the loose or not. Hicks reported that when travelers started vanishing along a certain path in Bilaspur, everyone assumed a tiger was to blame. A large government bounty was put on its head, and sportsmen went after it in droves, but though they found plenty of paw prints at scenes of "kills," none so much as caught sight of a tiger. Then one canny sportsman noticed that footprints at the scene of a "kill" were too small to have been the victim's, and that the victim had obviously been wealthy. His suspicions aroused, he inquired about previous victims. Sure enough, they had all been wealthy men. More, all had been strangers traveling alone, so no one knew what valuables they had been carrying. The sportsman went straight to the local police. They in turn wasted no time in arresting a yogi — a type of sadhu — who had taken up residence on the path just before the killings started. In his possession they found a stash of cash and jewelry, and two stuffed tiger's feet.[33]

Tracking the Weretiger

One evening in the 1920s, recalled Elwin, the Hindu master at a school near Karanjia, a man called Balinath, took a wealthy pupil into the jungle, stabbed him five times in the throat to make it look like the work of a tiger's claws, broke his wrists and ankles to prize off his gold bangles, and flung him into a cave known locally as a tiger's lair, where the poor boy eventually expired. The police were not fooled, however, and Balinath soon found himself being transported for life—presumably to the Andaman Islands, in the Bay of Bengal—only a legal technicality saving him from the gallows. But twelve years later, a reformed character, he was released; and naively came back to look for a teaching post in the same area. Understandably there was uproar on the day he returned, one of Elwin's colleagues, Mustapha Khan, even declaring his intention to barricade himself in his home that night if Balinath was allowed to stay. Elwin persuaded Balinath it would be best for everyone if he started a new life elsewhere, and put him on a train to distant Wardha (where he adopted an Untouchable boy, undertaking to care for him all his life). That night, when a leopard dug its way into the chicken-shed at Elwin's ashram and killed thirty of the thirty-one roosting birds, Elwin joked that the real culprit was Balinath "getting his hand in for fresh adventures." Noted the pastoral officer in his diary: "Mustapha Khan takes this quite seriously, and moves off with set and anxious face to strengthen his barricade."[34]

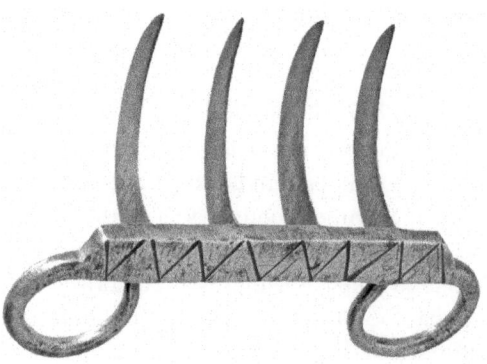

Figure 13. A simple *bagh nakh*, with the right number of "claws," namely four: some *bagh nakhs* wrongly have five "claws." In 1659, the great Maratha warrior Shivaji famously wielded a *bagh nakh* in combination with a *bichawa*, or curved dagger, to kill the giant Mogul general Afzal Khan (Mary Evans Picture Library/Interfoto Agentur).

Perhaps Balinath used a *bagh nakh*, or "tiger's claw" (Figure 13), the Indian equivalent of the metal "claws" used by African leopardmen to make it seem their victims were killed by leopards, and wielded by assassins and criminals all over India. (In the Malay world, the equivalent weapon was a *b'ladau*, noted Skeat and Blagden.[35])

Occasionally, individuals were wrongly accused in real cases of man-eating. Anderson recalled a villager in the

7. Beast People

Coimbatore district of southern India agreeing to sell a neighbor some land for 400 rupees. The two men set off on foot to the sub-registrar's office in the village of Talavadi to complete the paperwork. When they stopped by a stream for a rest, a tiger sprang out and snatched away the purchaser, together with his 400 rupees. The vendor ran the three miles to Talavadi, where on hearing his story the local police constable, with an eye to rapid promotion, arrested him on suspicion of murder and theft, without seeing the need even to inspect the scene. Luckily for the vendor a friend of Anderson's got wind of the matter and persuaded the constable's superiors to check the scene, where they found some rags, a few gnawed bones, and a number of 10-rupee notes fluttering in the breeze.[36]

Returning to Kahani, once word spread that Chinta was the true culprit, reported Sterndale, terrified farmers gave up guarding their crops, allowing deer and wild boar to raid them unmolested, while even the bravest shikaris became too scared to try to shoot the leopard, despite the government offering a 100-rupee reward. No one could blame them. One shikari sitting up in the dark on a field machan had a narrow escape when he fought it off by severing one of its toes with his knife, heard Sterndale, but at least one other shikari was not so lucky, the leopard plucking him from his platform and carrying him off into the night.

At dusk every villager retired behind closed doors, not daring to venture out again until daybreak. The head of each household stood watch all night, and every hour, recalled Sterndale, they would call out to one another, "*Jagte ho, biya? Jagte ho?*": "Are you awake, brothers? Are you awake?"

Not everyone exercised such sensible caution. One evening a government peon, a young Pathan, stopped at the home of the headman of Kahani, himself an honorable old Pathan. The night was sultry, so the young man elected to sleep on the veranda. Having himself once valiantly fought off the leopard by clouting it on the head with his ax, the headman implored his guest to come inside. But the youth scoffed at the idea, declaring that should the leopard dare to show its face he would draw his sword and slice it in half as easily as if it were a lemon. Reluctantly his host fastened the door. The young man strolled up and down the veranda, smoking his pipe, until around midnight he settled down to sleep. Two hours later came a terrible scream, and the peon was seen no more.

Related to Sterndale at a council of frightened shikaris at his camp in Kahani one morning, the fate of the peon confirmed how devilishly "cunning" the leopard had become. Now living solely on a diet of human flesh — or was it blood? — it was impossible to tempt with animal baits of any kind. A dummy of a man on a charpoy had been tried, but the killer had not been fooled. It was now so suspicious that having targeted someone — like the peon — it would lie in wait until the sound of their breathing assured it they were fast asleep. The dummy had made no sound at all, and for the same reason it would be useless to try to lure the leopard with a human corpse. What, then, could be done?

At this point the local *thanadar*, or police inspector — a pompous relative of the peon — made an extraordinary suggestion. If a live person was the only thing that would tempt the man-eater, then a live person should be tied out as bait. In Seoni jail he had two condemned murderers who would be perfect for the task. The sahib could sit over them, and being a renowned marksman was sure to drop the leopard dead at their feet. They would come to no harm, and could be hanged later.

(Such tactics were not entirely unknown. In 1874, according to American historian Philip McEldowney, the *Jabalpur Samachar* newspaper complained that Dr. French, the civil surgeon of the Bhandara district of the CPs, had tied an Indian to a tree "to serve as tiger bait." Fortunately for both parties the man had managed to free himself. Meanwhile, Baker reckoned this was the sort of thing an "Eastern potentate" might do.[37])

An astounded Sterndale told the thanadar his proposal was out of the question under British law. But the idea was a novel and ingenious one, he conceded, and he saw in it a way in which the brave inspector could himself become the hero of the district. (At this the man swelled with even greater than usual pride.) Everyone had heard him praise the district officer's aim, continued Sterndale. If he had such faith in it, why not volunteer himself as bait? There was no law against that.

At this there was much laughter, and the cornered inspector turned pale. But what if the sahib's normally trustworthy weapon failed to fire, he blustered? Who then would provide for his children? And after all, was not the leopard really a transformed Gond? Ordinarily, of course, he would willingly volunteer, but a wereleopard was quite another matter. Regretfully, then, he must decline the sahib's kind invitation.[38]

Finally, after much deliberation the council came up with a cunning plan. A fake camp-site would be set up, with a goat, muzzled to prevent it bleating, concealed in a wooden chest next to a dummy of a man on a charpoy. The leopard might then mistake the goat's breathing for that of a sleeping man.

Sterndale's followers made the dummy, which to everyone's amusement was nicknamed "the Thanadar Sahib." Allowing his followers all to sleep in one of his own large tents, with a fire kept burning at each door, and armed sentries posted front and back, Sterndale sat up over this ingenious lure several nights running. But the leopard was not deceived, and its tally of victims grew.

Eventually, though, the man-eater did meet its fate. The unwitting hero of the hour, wrote Sterndale, was a Gond called Kurria who eked a living as a village shikari by shooting game at night with an old matchlock left to him by his grandfather. When the man-eater began its reign he became too scared to hunt, until finally he and his wife, Sookea, had little left to eat. So one evening he nervously ventured out again. A few hours later he was hurrying home when he spotted what he thought was a wild boar in the shadows, knelt, and fired. At once there was a roar — clearly from a tiger or leopard — and Kurria fled. Back home Sookea berated him for his cowardice, thinking only of the chance to claim a handsome reward. Early the next morning she rounded up a dozen neighbors, all armed with axes and spears, and with Kurria in tow led them to the scene, where they found a dead leopard. Shot clean through the heart, it was instantly recognizable, from the ax wound on its head and its missing toe, as none other than the dreaded man-eater of Kahani.

Sookea danced with joy at the thought of all the bangles she could buy with "her share" of the 100-rupee reward, and as word of the leopard's demise spread Kurria was everywhere hailed as a hero. Initially, to Sookea's great annoyance, he could only claim the reward for an ordinary leopard, but the regulation six months later, when no further kills had been reported — confirming he had shot the man-eater — he was able to claim the full amount, and Sookea gleefully got her bangles. Sterndale, meanwhile, had bagged the leopard's skin: a beautifully glossy one that later adorned his drawing room back in London.[39]

Curiously, Forsyth wrote that the story of the Kahani killer being a wereleopard arose *after* its death, and that the culprit was said to be a devout local Hindu who, to chase away a real leopard blocking their path when he and his wife were returning from pilgrimage to the holy city of Varanasi (Benares) on the Ganges, transformed by swallowing a powder — ground root, presumably — and foolishly entrusted the antidote powder to her keeping.

Indeed, like the story of the reanimated, killer-aiding corpse, this is an archetypal tale in India, with various versions recorded over the years, including one by John Lockwood Kipling in 1891 and another by Hamel in 1915. Common to them all is the man's Actaeon-like retaining of his human brain in animal form. In Cambodia in the 1920s, American author Helen Churchill Candee — whose colorful life included surviving the *Titanic* (unlike Mr. K—), and treating Ernest Hemingway as a Red Cross nurse in Italy in World War I — heard that if you accidentally eat a certain kind of wild rice (that is, rice from the boundary between cultivation and the wild), you irreversibly turn into a tiger that is particularly sly for exactly the same reason.[40]

As well as being a neat cover story for a murder, then, the irreversible-transformation yarn is a semi–Faustian morality tale, for by temporarily acquiring the form and power of a tiger or leopard you risk losing your human form for ever. More, your grievance against all mankind, but particularly those among whom you once lived, makes you a menace of a man-eater condemned to be hunted for the rest of its days. Contrast this with O'Donnell's yarn, in which, solely out of a desire to exact murderous revenge, the Khond boy enters into a binding, fully Faustian pact knowing he will involuntarily transform every night for the rest of his life. This probably reflects O'Donnell's particular interest in medieval European werewolves, which supposedly made similar pacts with Satan.

Also common to almost all versions of the yarn is that the sorcerer relies at least partly on a secret root to work his magic. Most likely this was a *Dioscorea* tuber, or yam, of which there are many species in the tropics, all but one of which contain dioscorine, a toxic alkaloid only destroyed by repeated boiling. When swallowed or injected, dioscorine induces convulsions, anaesthesia, and adrenalin-rush. It has long been used worldwide by both healers and hunters — a small dose has a restorative effect, while a large one kills — and *Dioscorea hispida*, the Asiatic bitter yam, has long been used throughout Asia to make an intoxicating brew. Wessing noted that the Javanese for *Dioscorea hispida* is *gadung* or *gadong*, while Boomgaard observed that a verb from this word means to intoxicate, and therefore deceive, someone with the brew: as in macan gadongan, indeed.[41] One can guess this was also the root used by Baiga priests to induce tiger-spirit possession.

Another theme common to almost all versions of the yarn is the role of the wife as the bungling accomplice. One exception — in which roots also do not figure — was recorded by Sleeman, who recalled once talking with his friend the Rajah of Maihar, a small princely state between Jabalpur and Mirzapur in central India, about the large number of people being killed by tigers at the Kutra Pass there, and how best to tackle the problem.

"Nothing," said his friend, "could be more easy or more cheap than the destruction of these tigers, if they were of the ordinary sort; but the tigers that kill men by wholesale, as these do, are, you may be sure, men themselves converted into tigers by the force of their *science*; and such animals are of all the most unmanageable." And how did such men turn themselves into tigers, asked Sleeman? "Nothing," replied the rajah, "is more easy than this to persons who have once acquired the science; but how they learn it, or what it is, we unlettered men know not."

Once, he continued, there was a high priest of a large temple in the Maihar valley who habitually turned himself into a tiger, trusting a disciple to reverse the transformation by throwing a necklace over his tiger head the moment it happened. He had long given up the practice, "when he was one day seized with a violent desire to take his old form." All his old disciples had long since departed, so he asked a new one if he could trust him to stand by and throw the necklace. "Assuredly you may," came the reply; "such is my faith in you, and in the God we serve, that I fear nothing!" (Unfortunately, the rajah did not say who this god was, but the necklace strongly suggests Shiva.) At that the priest began to turn into a tiger. "The disciple stood trembling in every limb," said the rajah, "till he heard him give a roar that shook the whole edifice, when he fell flat upon his face, and dropped the necklace on the floor. The tiger bounded over him and out the door; and infested all the roads leading to the temple for many years afterwards."

Did the rajah think the priest was one of the tigers at the pass, asked Sleeman? Replied the rajah: "No, I do not; but I think that they may be all men who have become imbued with a little too much of the high priest's *science*— when men once acquire this science they can't help exercising it, though it be to their ruin and that of others." And if they were ordinary tigers, asked Sleeman? Then he would pay some Gonds 10 or 20 rupees to propitiate the victims' spirits guiding and protecting them: "If this is done, I pledge myself that the tigers will soon get killed themselves, or cease from killing men. If they do not, you may be quite sure that they are not ordinary tigers, but men turned into tigers, or that the Gonds have appropriated all you gave them to their own use."[42]

Sleeman also told a more typical version of the yarn, heard on a visit to the village of Deori in the Sagar district, north of the Narmada valley and the Mahadeo Hills, in 1831, when man-eating tigers were preying on a great many people in the heavily forested hills between Deori and the town of Sagar itself. The headman of Deori, Ram Chund Roo, assured him the killers "were neither more nor less than men turned into tigers—a thing which took place in the woods of central India much more often than people were aware of," the transformations being effected and reversed by eating secret roots. According to his own father, when he, Ram Chund Roo, was an infant, Rughoo, the family dhobi, or washerman, was one day "seized with a violent desire to ascertain what a man felt in the state of a tiger," so he dug up the necessary roots and foolishly asked his wife to stand by with the reversal one.... "Poor old Rughoo took to the woods," said the headman, "and there ate a good many of his old friends from the neighbouring villages; but he was at last shot and recognized from the circumstance of his *having no tail*." This, stressed Ram Chund Roo, was the only thing that distinguished a weretiger from an ordinary tiger.

Commented Sleeman: "How my friend had satisfied himself of the truth of this story I know not, but he religiously believes it, and so do all his attendants and mine; and out of a population of thirty thousand people in the town of Saugor [Sagar], not one would doubt the story of the washerman if he heard it."[43]

A dhobi was also accused in an unusual spate of killings that took place in the Jhansi district of central India in 1928, as witnessed by Mrs. E. Minshull, the district officer's wife. Baby in tow, the Minshulls were on their annual dry-season tour when they arrived one morning on the outskirts of the village of Bansi on the Shahzad River. As usual their servants had traveled ahead overnight to make camp and prepare breakfast, but as they approached they were annoyed to see no tents pitched. Indeed, the place was in pandemonium, and they were soon surrounded by excited servants, their *khitmagar*, or table servant, almost throwing himself under the wheels of their automobile before crying out: "Huzoor, let us be gone from this accursed place as soon as possible. A fearful demon haunts it and has killed hundreds of villagers. Yet no one has seen it, and down in the village they tell us that it is the spirit of a dhobi, who lived on the banks of the river. An evil spirit entered into him and he slew his wife and devoured her. Now he wanders through the jungle with matted hair and nails like claws, going on all fours and destroying all he meets."

The Minshulls sent a more level-headed servant into Bansi to find out what was really going on, and this man confirmed that an unidentified creature — whether a tiger or a leopard, no one could say — had indeed been killing and eating people in the area, including twenty in Bansi alone, for some while. Only the night before, it had entered a neighboring village, all the terrified inhabitants of which, bar the chowkidar, had barricaded themselves indoors: in the morning the night-watchman's decapitated body was found near his bed.

That night the Minshulls ordered fires be kept burning around their camp, and slept with loaded rifles by their beds. The next day they joined forces with the district police superintendent, and there followed days of chasing after news of fresh kills—each up to fifty miles apart—and nights of sitting up in tree machans over live baits. "But still the creature killed and killed," recalled Mrs. Minshull, "and gave no sign by which it could be identified."

One day an old man arrived from a village twenty miles distant, having run almost all the way. When he got his breath back he said that early the previous evening he was sitting by the fire in his hut when "an enormous yellow monster," as he described it, "with a great ruff of yellow hair," stole in through the open door and seized his grandson from his bed. The man was coy about what had happened next, but it transpired that with great courage and presence of mind he had shoved a flaming faggot in the creature's face, causing it to drop the boy and flee. Miraculously, the boy was unharmed.

Then the creature seized a woman in a nearby field, and at last they could examine the undisturbed scene of a fresh kill. There they found the paw prints of a big cat that was lame in one foreleg, but clearly was no tiger or leopard. No one was sure, but in light of the old man's testimony they all agreed that, however unlikely it seemed, the creature must be a lion.

Whatever it was, despite the government's offer of a 500-rupee reward, the Minshull party and others all failed to bag it, and it continued to kill again and again; until one day the killings stopped, and it was heard of no more.[44]

The killer could not have been an Asiatic lion, an animal restricted to the Gir forest of the Kathiawar Peninsula, hundreds of miles to the southwest, but it could, thought sportsman Stanley Jepson, have been an African one. With an eye to spicing up the local shikar, some years earlier the maharajah of the neighboring princely state of Gwalior had taken it into his head to import four pairs of African lions and release them into the wild, where they and their offspring took to man-eating. The Jhansi killer may have been one of these that strayed into the district before returning to Gwalior where, like the others, it was eventually shot.[45]

Dhobis are Untouchables, and have long been widely associated with sorcery, largely, perhaps, because they work with water, and because of the purificatory, transformative nature of the work they do for people of higher caste. Ram Chund Roo also told Sleeman they were renowned drunkards, and alcohol abuse may be as good a way as any other of contacting the spirit world. After all, Baiga priests imbibed pretty freely of the mahua spirit when preparing to summon spirit tigers. Writing in Edgar Thurston and K. Rangachari's 1909 *Castes and Tribes of Southern India*, L. K. Anantha Krishna Iyer described the Velans, or Mannans—a largely dhobi caste—of Cochin, now Kochi, in Kerala, in southwest India, as "devil-dancers, sorcerers and quack doctors," animists who worship such "demoniacal" deities as Chathan, a son or incarnation of Shiva, and Bhagavathi, meaning Durga or Kali. In the princely state of Travancore in southwest India, added N. Subramani Iyer, Velans were said to be descended from Shiva, who they worshipped in preference to all other deities.[46]

Anantha Krishna Iyer recorded two shapeshifting prescriptions issued by Velans. One is: "Take the head of a dog and burn it, and plant on it vellakutti plant [I have been unable to identify this]. Burn camphor and frankincense, and adore it. Then pluck the root. Mix it with the milk of a dog and the bones of a cat. A mark made with the mixture on the forehead will enable any person to assume the figure of any animal he thinks of." The other is: "Before a stick of the Malankara plant [the emetic nut tree (*Catunaregam spinosa*)], worship with a lighted wick and incense. Then chant the Sakti [Shakti, or empowering] mantram 101 times, and mutter the mantram to give life at the bottom. Watch carefully which way the stick inclines. Proceed to the south of the stick, and pluck the whiskers of a live tiger, and make with them a ball of the veerali silk [as used to cover Keralan warriors killed fighting heroically], string it with silk, and enclose it within the ear. Stand on the palms of the hand to attain the disguise of a tiger, and, with the stick in hand, think of a cat, white bull, or other animal. Then you will, in the eyes of others, appear as such."[47]

Plucking the whiskers of a live tiger sounds about as easy as milking a tigress.

In 1957, Powell reported a version of the irreversible-transformation yarn told to him by two old shikari friends in Balaghat, one a Baiga, the other a Gond. To slay a real man-eater of the neighborhood, they said, a banyan, or moneylender, transformed himself into a tiger by swallowing a powder given to him by a sadhu for the task. Sadly he entrusted the antidote to his pretty young wife. After killing and eating her he became "the most dreaded man-eater in human memory," claiming thousands of lives.[48]

Anyone in colonial India would have smiled knowingly on hearing this version of the tale, such were the reputations of banyans and sadhus alike. Russell and Hira Lal noted that for centuries the village banyan was a respected figure with a reputation for fairness. Lending money on little security, and being flexible about repayments, he was a key part of village life. But by the 1850s, many banyans were installed by outside firms, had no interest in the local community, and insisted on prompt repayments. And under British law a banyan could take to court any peasant farmer who failed to meet his payments, which often resulted in the banyan becoming the peasant's landlord, and the peasant the banyan's permanently indebted tenant with no rights to the land at all: this after centuries of peasants having a hereditary right to farm, if not own, the same land as their forebears. As most peasants were illiterate, it was easy for grasping banyans to exploit them when drawing up the terms of a loan.[49]

Tribespeople who became settled farmers were routinely swindled this way. One morning soon after his appointment to Bilaspur, Best rode out to visit the Baiga village of Surhi in the Lurmi Range there. The Surhi Baigas were almost unique in tilling the land, growing rice and tobacco. On his approach, Best spied one hiding in the undergrowth. What was he scared of, he asked; did he expect the sahib to eat him? Laughing, the man said he was hiding from the banyan. Two years earlier, he explained, the banyan had lent them the money to buy the tobacco seed, at 200 percent interest. They had managed to pay off only the capital and a quarter of the interest, and now it was harvest time the banyan had turned up intent on seizing the crop and having them all thrown in jail.

Best made camp under the village peepul tree, breakfasted, then sent for the banyan. "He was fat," he observed, "with many wrinkles on his bare belly, and an ingratiating smirk on his oily face." When questioning of this charming character revealed he was camping out nearby, Best called his bluff by declaring he was committing three serious offenses under the Forest Act — settling in a forest village without permission, taking firewood without authorization, and burning said firewood in a fire-protected area — for any one of which he could be jailed for six months and fined 500 rupees. At this the banyan sweated profusely, threw himself on the ground, and pleaded for mercy. Best gave him an hour to clear out.[50] Small wonder a banyan might be put in the frame when a man-eater was around.

As for sadhus, like banyans they are outsiders, but more, they are strangers in the full sense of the word. In their search for enlightenment, many go *sannyasi,* renouncing all possessions and pleasures, wandering

India as beggars, and spending years traveling to distant shrines, or walking the banks of the Narmada from sea to source and back again. They may stay on the fringes of a community for a while before moving on, and may even be welcomed; but only for as long as all is well in the neighborhood.

Sadhus perform magic rites to contact the gods, with yogis, noted Russell and Hira Lal, even practicing such forms of mortification as holding one arm aloft for year after year until it withers away in order to acquire mystical knowledge and power (though, confusingly, many colonials applied the name "yogi" to any kind of sadhu). Most are devotees of Shiva and cast themselves in his image, going about naked or semi-naked, their bodies smeared with ash, with long hair and nails, tridents, necklaces, and strings of beads. At one time, many carried leopard or tiger skins to sit and sleep on, just like Shiva.[51]

A few Shivaite sadhus, the Aghoris, use the tops of human skulls as begging bowls, and live on burning-ghats— the haunts of witches and evil spirits— where they smear themselves with the ashes of burnt corpses. As an expression of the doctrine that, everything being full of Brahma, one thing is as pure as another, they occasionally eat shreds of flesh that survive the funeral flames, not to mention excrement. Like the Kapalika medieval sect who worshipped Bhairava and Chamunda, a particularly horrible form of Durga, they have long been rumored to perform human sacrifice, and to be great sorcerers. "It is believed that an Aghori can at will assume the shapes of a bird, an animal or a fish," wrote Russell and Hira Lal, "and that he can bring back to life a corpse of which he has eaten a part."[52]

Most sadhus are genuine, but a few have always been at best fakes who perform magic tricks for alms, and at worst, like the yogi cited by Hicks, outright criminals. Like banyans, some were also far from penniless in colonial times. On February 1, 1935, Elwin wrote in his diary: "Very important ascetic, who for years has renounced the world and its possessions, murdered at Amarkantak and his private store of cash, estimated at 10,000 rupees, stolen. Only really wealthy people in this district appear to be the *sadhus*."[53]

It has long been fashionable for wealthy Indians to go sannyasi at some point to redeem themselves, but other sadhus accumulated their wealth fraudulently. Whenever man-eaters infested a road in Bengal, reported Burton, sooner or later a fraudulent yogi would turn up, erect a hut, and sit in it all day. For a fee he would then sacrifice a cockerel and pray to ensure the safe passage of travelers. Should one then still fall victim to a tiger he would excuse himself by saying they had obviously been too great a sinner to be saved.[54]

7. Beast People

Sanderson recalled a Shivaite sadhu turning up in Mysore with two "attendants," who widely publicized their "guru" while the man himself sat chanting. He soon had hundreds of disciples. Saying the gods had granted him supernatural powers, and that he was planning a week-long trip to the underworld, his attendants had some disciples dig a pit and cover it with an earthwork shrine, topped by a clay bull, an emblem of Shiva. Meanwhile, close by they built a hut for him to sit in. On the appointed day he entered the shrine, but despite it being carefully guarded, in the days that followed he popped up all over the place. Wrote Sanderson: "The fact that he was wandering about the country ... does not appear to have struck any of his believers as a departure from his original undertaking." At the end of the week he calmly emerged "shaking the soil from his matted locks." The crowd hailed him as a god, sat him on the bull, and threw coins at his feet. Wrote Sanderson: "Some gave as much as thirty or forty rupees; and a sum of upwards of £200 was thus contributed." Leaving his attendants to fill in the pit, the sadhu then went around "levying further contributions with a cupidity scarcely consistent with his unworldly character." The next rainy season, long after he and his accomplices had left, the shrine collapsed, revealing a tunnel to the hut.[55]

Real or fake, a sadhu who just happened to be in the vicinity was an obvious candidate for blame when a man-eater was on the loose, as we shall see in the case of perhaps the most notorious man-eater in Indian history. And though I have found no evidence of it, it seems likely that, like some Javanese beggars, some sadhus must have extorted money by threatening people with their supposed shapeshifting powers. Mitra certainly noted it as a general trait of weretigers in India.[56]

In reality, sadhus often fell victim to man-eaters themselves, as shown by the fate of so many at Amarkantak. In large part this was because of their solitary, wandering lifestyle. Forsyth recalled finding what little remained of a lone sadhu in an isolated nullah in Mandla in January 1863: his trident, human skull begging bowl, and hookah, a few shreds of cloth, some tresses of his long, matted hair, and a handful of crunched-up bones.[57]

But it was also due to a mistaken conviction that no animal could harm them, something that also applied to fakirs, or Muslim ascetics (though, confusingly, as Russell and Hira Lal pointed out, in northern India sadhus were also known as fakirs). Other people believed it too. "You may hear in any Hindu village," wrote John Lockwood Kipling, "of jogis [sic] to whom the cruel beasts are as lapdogs. In the native newspapers, as in popular talk, cases are reported in complete good faith where a Rajah out hunting is endangered by a mad wild elephant or a ferocious tiger. At the critical moment the jogi appears and orders the obedient beast away." Likewise, it was said in Sundarbans, noted Mitra, that if cornered by a tiger, all you had to do was shout for help from one of the many yogis wandering around. A yogi would then walk up to the tiger, pat it on the back, and tell it to go away.[58]

Wrote Eardley-Wilmot: "I often questioned wandering 'fakirs,' holy hermits and other individuals leading the simple and solitary life in wild places in the forests, as to their attitude to the animal life around them; and they gave me to understand that power over the beasts of the field was one of the earliest results of their self-imposed austerities. None expressed any fear of attack from their wild neighbours, and one even offered to call them up if I would promise not to shoot them on arrival." Kipling recalled a fakir in Lahore thrusting his arm through the bars of a tiger's cage to prove his power over the animal. The tiger tore the limb from his body.[59]

8
Beast Master: The White Sadhu and the Ultimate Terror

"The tiger is, as a rule, a gentleman. The panther, on the other hand, is a bounder."—Alexander Glasfurd, *Leaves from an Indian Jungle*, 1903, 55

For eight long years, from 1918 to 1926, a single male leopard terrorized an area of some 500 square miles around the Himalayan foothills village of Rudraprayag in the Garhwal district of Kumaon province — a vast range for any man-eater — putting its 50,000 residents in nightly fear of their lives, and terrifying the 60,000 Hindu pilgrims who annually passed through on their way to and from paying homage at the mountain shrines of Badarinath and Kedarnath. ("Rudraprayag" means "Rudra's confluence," Rudra, or "storm," being one of eight aspects of Bhairava.) Each evening, everyone living in the area rushed home and barricaded themselves in, while footsore pilgrims forgot their weariness and hastened to roadside shelters. When night fell, silence descended over the hills, no one daring to move or even speak lest the man-eater be prowling nearby. People in the vicinity of Rudraprayag had good reason to be afraid.

One night a 14-year-old goatherd — an orphan and an Untouchable — bedded down as usual in his charges' windowless pen. Mindful of the leopard, his master jammed a piece of wood through the hasp of the door. The next day he found what little was left of the lad in a nearby ravine. The leopard had clawed at the door until the piece of wood had fallen out then, ignoring the panicking goats, had made straight for the boy cowering at the back of the pen.

Another night, a buffalo herdsman, his 12-year-old daughter, and his brother were sleeping as usual in the open, confident of the protection of the herd. During the night, clanging buffalo bells and frightened snorts woke the men. Leaving the girl still sleeping soundly, they went to investigate. On their return just minutes later they found only great splashes of blood.

Then there was the night two men sat side by side in a house sharing a hookah in the dark. The door, just out of view, was closed but not fastened. The man whose house it was had just passed the pipe to his friend when it fell to the floor, scattering charcoal and tobacco. Admonishing him for his clumsiness, he leant forward to clear up the mess; and saw the silhouette of the leopard slipping out the door, his friend dangling dead in its jaws.

On another occasion a man slept in the outer room of his house, while his sick wife slept in the inner room between two friends who were nursing her. The leopard stole in through a narrow window — without knocking over the brass jug almost filling it — skirted the man, slipped into the inner room and clamped its jaws around the sick woman's throat. Only when it knocked over the jug as it tried to drag her corpse out through the window did the rest of the household awake, at which it dropped her and fled. (The husband of one of the friends said it was lucky the leopard chose her as she would likely have died soon anyway.)

Finally, the leopard once pushed open the door of a house and seized the woman sleeping inside by one leg. As it tried to drag her, screaming, outside, the door swung shut on the limb. With one tug the leopard tore it from her body.[1]

All told, the leopard claimed an officially recorded 125 victims. It seems it acquired its taste for human flesh when the global Spanish flu pandemic of 1918 swept through India: a time, recalled Best, known as the *garbar*, or confusion. Most Garhwalis could not afford to dispose of their mounting dead in the traditional manner, by cremating them and scattering their ashes in the Ganges or one of its tributaries. Instead they placed embers in their mouths and tipped them into the nearest nullah. The ready supply of corpses provided a feast for scavengers, including leopards; and when it ended, this particular leopard may have taken to killing people because it had become accustomed to eating nothing else.[2]

Whatever the cause of its diet, not only was the man-eater deadly determined, and extraordinarily stealthy, it seemed to be charmed. Despite the government offering rewards — wildly rumored to total 10,000 rupees plus the ownership of two villages — granting more than 300 extra gun licenses on top of the 4,000 already held in Garhwal, retaining shikaris on generous wages, allowing soldiers to pursue the leopard on leave, and appealing for sportsmen from all over India to try for it, the best anyone did was to shoot it in the foot.

At the same time numerous drop-door traps, baited with live goats, were set for the man-eater, but even when a leopard everyone agreed was the man-eater was caught in one no Hindu dared kill it for fear its victims' spirits would haunt them in revenge. So a Christian Indian from thirty miles away was sent for to do the deed; and by the time he arrived the leopard had dug itself free.

Another time the man-eater was chased into a cave, which was then sealed with rocks and thorn bushes. Five days later a local dignitary turned up, scornfully declared the cave empty, and ordered it to be unsealed. Out shot the leopard, straight through a crowd of 500 startled onlookers.

The half-eaten remains of some victims were laced with arsenic and strychnine, and left where they lay. The leopard came back and happily ate from them, apparently suffering no ill effects whatsoever.

Finally, in the fall of 1925, by which time the leopard was known to have killed more than 100 people, Bill "Ibby" Ibbotson, deputy commissioner of Garhwal, turned for help to an old friend in neighboring Naini Tal district: Jim Corbett.[3]

Of Irish descent, Corbett was born and raised in the hill station of Naini Tal, where his father was postmaster. Shooting was his passion, but he was never a brash sportsman in the typical mold. Like so many other domiciled sahibs and memsahibs, he was fiercely patriotic — some might say absurdly so, given the way his kind were looked down on by most other colonials — but he was a shy, modest man with a deep affection for India and its people.[4]

Above all he loved Kumaon. From the age of seventeen he worked for more than twenty years for the Bengal & North-Western Railway in Bihar, but he made the week-long journey home every year on leave. Then in 1917 he recruited 500 Kumaonis, served with them on the Western Front, and, according to two of his biographers, Martin Booth and Durga Kala, at the end of hostilities brought all but one safely home.[5]

After a stint fighting Afghans on the North-West Frontier he took over a combined hardware store and house agency in Naini Tal, building it into a flourishing business. With a profitable coffee farm in Tanganyika (now Tanzania), this left him ample time for shikar.

Over the years, though, Corbett became increasingly conservation-minded, and eventually, partly inspired by the pioneering work of UPs forest officer Frederick Champion, took to photographing and filming animals instead of killing them: as did many sportsmen when technological advances allowed it, and when, at the end of a long career, they had had their fill of slaughter, and could see the forests and game vanishing at an ever increasing rate. Such converts, who included Hanley, Anderson, Colonel Sir James Sleeman (William Sleeman's grandson), and Theodore Hubback, the first person to film a Malay Peninsula tiger, came to value a photograph or film as a more fitting and lasting trophy than a moldy old head or skin. Corbett was instrumental in setting up India's first tiger reserve — Hailey (later Ramganga) National Park, in Kumaon — and his standing as a pioneer conservationist was recognized after his death; first in 1957, when the park was renamed Corbett National Park, then again in 1968, when the Indo-Chinese race of tiger was named *Panthera tigris corbetti*.

Corbett began his career against Kumaon's killer big cats in quite spectacular fashion, accounting in 1907, at his very first attempt at shooting a man-eater, for one of the most notorious tigers in history: the Champawat tigress, a veritable shaitan blamed for the deaths of no fewer than 436 people. Three years later he followed up that triumph by bagging two more infamous "devils": the Muktesar tigress, and the Panar leopard. The former was credited with a modest haul of twenty-four victims, but the latter — like the Rudraprayag killer, a male — was said to have killed more than 400 people in the five years it was active around the hill station of Almora in the district of the same name: in 1935, Burke, the district's forest officer at the time, and his daughter Norah put the total at more than 500.[6]

Questions were even asked in the House of Commons about this creature, but, outside Almora, in India it never achieved the notoriety of the Rudraprayag man-eater. Corbett reckoned this was because it operated in a much remoter area, far from the pilgrim trail. Inside Almora it was a different story. As the Burkes remarked, despite the government offering a 500-rupee reward for it, the people of Almora "came at last to believe that it was useless to try and kill him because he must be a devil in a panther's skin." Squads of Gurkhas were sent in, to no avail. Two Indian soldiers once tried to bag the killer, hiding in rocks near a victim's remains. The leopard carried one of them off.[7]

When the government asked Corbett to hunt it, he first visited the deputy commissioner of Almora, a man called Stiffe, to find out all he could about the animal. "He kindly invited me to lunch," recalled the sportsman, "provided me with maps, and then gave me a bit of a jolt when wishing me goodbye by asking me if I had considered all the risks and prepared for them by making my will."[8]

Undeterred, Corbett eventually shot the leopard while sitting in a tree, shivering with malaria, over a tethered goat; though not before it had spent most of the night trying to drag him from his perch.[9]

Each time Corbett hunted a man-eater, Kumaonis marveled at his bravery and extraordinary jungle craft, which included a keen sixth sense for danger. Fittingly enough for a wiry, abstemious, almost ascetic man who even when at home in the dry winter season preferred to sleep outside in his tent, his successes against man-eaters earned him an enduring reputation as a sadhu whose powers were greater than those of any evil spirit. As he himself modestly admitted: "However little I merit it, the people of our hills credit me with supernatural powers where man-eaters are concerned." When he embarked on his hunt for the Chowgarh tigress, one survivor of an attack told him: "Do not be deceived into thinking it is a tiger, for it is no tiger but an evil spirit who, when it craves for human flesh and blood, takes on for a little while the semblance of a tiger. But they say you are a sadhu, sahib, and the spirits that guard sadhus are more powerful than this evil spirit."[10]

Above all, Kumaonis appreciated that he hunted man-eaters to save lives and livelihoods, not to get coolies back to work. As he remarked of his 1938 triumph over the Thak tigress, when he acceded to the Forest Department's request to hunt it he did so more in the interests of the local people than in those of the contractors signed up to hire them to fell trees in the area.[11]

As for the Rudraprayag leopard, Corbett had not gone after it before 1925 because he had assumed that sooner or later someone else would bag it. Then came Ibbotson's letter. A confirmed bachelor, Corbett lived with his devoted spinster sister Maggie, who pleaded with him not to go. After all, he had not hunted a man-eater for fifteen years and was now in his fifties. But Corbett fancied his chances and duly set out for Garhwal with his faithful servant Madho Singh and six bearers. First, though, as always he insisted the field be closed to all other parties to give him a clear run at the leopard (not that anyone else was willing to go after it by that time). And as always, to emphasize he was no bounty hunter, he made it clear he would accept no reward should he succeed in killing it.[12]

On arriving at his base on the pilgrim road at Rudraprayag—the one-room Inspection Bungalow, with outbuildings for his men—Corbett bought two goats and tethered one on the road and the other on a path bearing the man-eater's paw prints. The leopard killed the goat on the path that night, so the following afternoon he sat up in a tree in case it returned. By nightfall it had not shown, so he warily headed back. His caution was well founded, for in the morning a trail of paw prints showed that the leopard had stalked him all the way.

The next day came news that the leopard had killed a woman from a village three miles away, 4,000 feet up the hill. Arriving drenched with sweat from the climb, Corbett found a harrowing scene. The woman's half-eaten remains lay in a ravine, her belly ripped open to expose her unborn son. Grim-faced, he set to work. A traditional Kumaoni haystack in a walnut tree would serve as a machan. A flashlight he had ordered to fix to his rifle had not arrived yet so he placed a white stone as a marker on the opposite side to where the leopard had lain down to feed. As a back-up he lashed his shotgun and spare rifle to bamboo stakes on one side of the path into the ravine and stretched silk fishing line from their triggers to stakes on the other side.

That night, no sooner had darkness fallen than Corbett heard the leopard enter the ravine, having somehow avoided the gun-trap, and begin scratching about in the straw below him. Just then a terrific storm blew over, soaking him to the skin and chilling him to the bone while the leopard lay warm and dry under the tree. Through the deluge he saw a lantern weaving toward the village. Later he learned its brave bearer had marched thirty miles to bring him his flashlight.

When the storm passed, Corbett heard the leopard resume its meal, but could see nothing: evidently a puddle had formed on the side where it had lain before, so now it was on the other side, obscuring the marker. Over the next few hours it moved repeatedly off the kill but, maddeningly, each time his rifle was trained on the stone it returned just as the weight of the weapon forced him to lower the barrel and rest his aching arms. Finally he chanced a shot in the gloom. The leopard fled, avoiding the trap again, but a tuft of hair on the ground the next morning showed that he had grazed its neck. It had had another lucky escape.

Corbett now made a tough decision. The man-eater killed on both sides of the Alaknanda River and was now on the east side, where the terrain favored the sportsman. By closing the two suspension bridges—one at Rudraprayag, where the Mandakini River joins the Alaknanda to form the Ganges, and one twelve miles upstream, at Chatwapipal—he could halve the search area at a stroke. Everyone on the west side would be safe, but everyone on the east side at the leopard's mercy. After much soul searching he ordered the bridges to be blocked off at night with thorn bushes. Such was the faith the Garhwalis had in him — a faith he found "as touching as it was embarrassing"— not one person on the east side protested.[13]

Before having the Rudraprayag bridge closed, however, Corbett spent twenty consecutive nights lying alone with his rifle on a windswept, ant-infested ledge atop one of the bridge's stone towers, high above the raging, icy river: "not," he noted, "that the temperature of the water would have been of any interest after a fall of sixty feet on to sharp and jagged rocks."[14]

The leopard never showed, but on the second day a white-robed Christian Indian with long hair and straggling beard crossed the bridge deep in prayer and carrying a six-foot silver cross. He appeared again the next day and told Corbett he had come from a distant land to free Garhwal from the evil spirit tormenting them. To this end he set about making an effigy of a tiger out of bamboo, paper, string, and cloth. After several days' work a night of heavy rain left the near-finished item in tatters, but he cheerfully began again from scratch. Finally the figure — "about the size of a horse, and resembling no known animal," wrote Corbett — was ready for a *tamasha*, or show.[15]

Lashed by its feet to a pole like a real tiger, and accompanied by more than 100 Hindus banging gongs and blowing trumpets, the effigy was carried down to the Ganges, where its maker knelt in the sand and beseeched the evil spirit to enter it. The crowd then launched it into the river, throwing handfuls of sweets and flowers after it to speed it on its way to the sea.

After Corbett's vigil on the bridge, Ibbotson arrived to help, and promptly commandeered the bungalow compound, so Corbett and his men had to encamp across the road and fence themselves in as best they could with thorn bushes. A prickly-pear tree hampered the erection of Corbett's tent, but halfway through having it cut down he decided it provided much-needed shade and ordered his men just to lop off the offending branches. Only when his tent was up did it occur to him the tree also provided an easy way in for the man-eater, but by then it was too late to do anything about it. Sure enough, as Madho Singh snored on his back outside the tent that night the sound of the leopard climbing the tree woke Corbett from a fitful sleep. Clearly the tamasha had not worked.

Corbett snatched up his rifle, ready-loaded on his bed, and was just pulling on his slippers when there came a loud crack followed by a yell of "Bagh! Bagh!" Rushing out, he saw the leopard bound away into the night.

When Madho Singh calmed down he said that on hearing the branch break he had opened his eyes and looked straight into the face of the man-eater preparing to pounce.

Corbett had the tree cut down the next day.

A few days later came word that the leopard had killed a cow in a nearby village. That night Corbett and Ibbotson sat up over the carcass in a two-tiered machan they craftily made to look like a local haystack. After a while Corbett, on the lower tier, heard the leopard arrive and begin crawling towards the dead cow. But at the critical moment Ibbotson, unawares, moved his legs to ease them of cramp and scared it away. Luck had saved it again.

The pair next tried a gin-trap: the most fearsome one Corbett had ever seen, with three-inch teeth and springs so powerful it took two men to prize it open. An hour after dark on the first night they set this devilish device, near another killed cow, a leopard trod on the plate, and the jaws smashed shut on its leg. After putting the roaring creature out of its misery Corbett examined it and was sure it was not the man-eater. But within minutes it was surrounded by a dancing crowd who lashed it to a pole and carried it triumphantly from door to door. Later that night Corbett was proved right when the real man-eater struck again, killing a young woman when she nipped out to answer a call of nature.

The next day a guide led Corbett and Ibbotson to a thickly wooded ravine where the leopard had left the woman after carrying her more than half a mile. She lay face down, her hands by her sides and stripped of all clothing, with all the skin from the soles of her feet to her neck rasped away by the killer's rough tongue, and two gaping wounds in her back and buttocks where it had eaten several pounds of flesh. That afternoon the two men settled in a nearby tree.

At sunset a *kakar*—a muntjac, or barking deer—dashed barking down the hillside; the leopard was back. But by the time they heard it approaching, night had fallen and they could see nothing. Worse, they were in a direct line between it and the corpse, so it was bound to see them. Sure enough, they soon heard the leopard heading straight for them. It was time to leave.

As they hurried away, Ibbotson slipped and smashed their only lamp on a rock. Stumbling blindly across unfamiliar ground, they expected the man-eater to pounce at any moment. But they reached the house they had arranged to stay in and had just shut the door when a dog began yelping in terror outside. Looking out they saw it crouching, hair on end, its gaze following the retreating leopard in the dark. They had made it just in time.

8. Beast Master

When the leopard killed another cow a few miles away, and Corbett saw it had fed from the carcass with its front paws resting between the cow's legs, he buried the gin-trap there under a layer of big green leaves with a sprinkling of earth, dry leaves, and splinters of bone over them to make the scene look totally undisturbed. At sundown, as he walked alone to the local headman's house — Ibbotson having returned to his other duties — he became aware the man-eater was stalking him once more. Later that night, and not for the first time, the headman heard it claw at his door. In the morning Corbett found it had returned to the cow and continued its meal with its front paws resting safely on the levers of the trap.[16]

Another morning Corbett met an old Banjara on the pilgrim road. As he sat smoking Corbett's cigarettes, he told the sahib the following story. When his father was a young man and a leopard was terrorizing the neighborhood, the village headmen of the area gathered to discuss what to do. An old man whose grandson had been killed only the night before stood up and said matter-of-factly that the killer was "one from among their own community who, when he craved for human flesh and blood, assumed the semblance of a leopard," and that he strongly suspected a fat sadhu who lived in a hut near some temple ruins. At this there was uproar. Some said the old man was demented, others that he was surely right. Had the killings not started just after the sadhu arrived? And did not the sadhu always sleep all day in the sun after a killing the night before? (And how else, I might add, did he come to be so fat?)

When the hubbub died down the headmen ordered a close watch be kept on the suspect. For two nights he stayed in his hut and no killings were reported, but on the third night the men on watch, including the Banjara's father, saw him slip out into the dark. A few hours later they heard a terrible scream come from the direction of a charcoal-burner's hut up the hill; and at dawn they saw the sadhu hurrying home, his hands and mouth dripping blood. As soon as he was inside, the men rushed over and fastened the door with a chain, piled straw around the hut, and set it ablaze, incinerating its occupant. "From that day," the Banjara assured Corbett, "the killing stopped."

Likewise, he said, in the present case it was only a matter of time before suspicion fell on one of the many sadhus in the area; and when it did, the same method of dealing with him would be employed. Until then the killings would continue, and Corbett was wasting his time.

"Suspicion" had in fact already fallen on a sadhu — a man called Dasjulapatty, from the village of Kothki — earlier in the Rudraprayag man-eater's career. A mob were about to lynch him when Ibbotson's predecessor turned up. No doubt, he said, they had the right man, but in the interests of justice he proposed he be locked up until his guilt could be proven. Now calm, the mob agreed and handed him over. A week later, when the leopard struck again, they raised no objection to his immediate release. This time, they said, they had got the wrong man; but next time there would be no mistake.[17]

In recalling this story—which shows the supposed metamorphosis was physical, since walls are no barrier to a spiritual shapeshifter—Corbett identified the canny deputy commissioner in question as Philip Mason, who in fact held the post much later, from 1936 to 1939, not having joined the Indian Civil Service until 1928. Writing as Philip Woodruff, Mason incorporated the tale in his novel *The Wild Sweet Witch*; published in 1947, a year before Corbett's account (and reprinted under his real name in 1989). In his foreword, Mason wrote that the deputy commissioner of the day really did take a man into custody to prove he was not the leopard, but that his own knowledge of this was based on hearsay.

According to Corbett, Garhwalis said all man-eaters were sadhus, who killed out of a lust for human flesh and blood. With so many sadhus passing through Garhwal on pilgrimage, this is perhaps understandable. In Naini Tal and Almora, however, he wrote, people blamed Bhoksas, giving robbery as the motive (the actual man-eating presumably being just a cover, or merely a bonus). But he made no mention of anyone claiming the culprit to be a Bhoksa when telling the story of the Panar leopard. In 1896, Crooke reported that Bhoksas were said to turn into man-eaters to avenge enemies.[18]

Anyway, if the Rudraprayag leopard really was a sadhu, it was not averse to preying on its own kind. One evening twenty pilgrims struggling up to Badarinath stopped at a one-room store. Exhausted, they asked the owner if they could sleep on his doorstep. The man refused, saying it would mean certain death, as he had often heard the man-eater prowling around. He urged them to hurry on to the shelter four miles up the road. But just then a sadhu arrived and declared that if the owner made room inside for the women, he would sleep outside with the men, and—shades of Sterndale's Pathan peon here—if any leopard, man-eater or otherwise, dared to molest them "he would take it by the mouth and tear it in half." The owner reluctantly consented and all settled down to sleep. The next morning the sadhu was gone. Following a trail of blood they found him sprawled across a low wall, his lower half eaten away. Ibbotson organized a beat of the area that afternoon, assembling no fewer than 2,000 men. But the leopard was long gone.[19]

After ten weeks of hunting the Rudraprayag man-eater, a dispirited Corbett returned home for the winter, exhausted as much by the strain on his nerves as by trailing miles every day from one remote village to the next and sitting up over kills night after cold night. On the way home — itself a tiring trek taking many days — a snake crossed his path: always an ill omen. "There goes the spirit that has been responsible for your failure," said Madho Singh.[20]

Corbett had always greatly respected the superstitious beliefs of Kumaonis, and now was inclined to agree the leopard possessed "devilish cunning." Many times during his vigils, he admitted, he pictured it as having the head of a fiend: "A fiend who, while watching me through the long night hours, rocked and rolled with silent fiendish laughter at my vain attempts to outwit him, and licked his lips in anticipation of the time when, finding me off my guard for one brief moment, he would get the opportunity he was waiting for, of burying his teeth in my throat."[21]

Though no one else was willing to hunt the leopard, Ibbotson withdrew the government reward while Corbett was away. In a letter to *The Pioneer* newspaper he said that by leaving the leopard undisturbed "and allowed, as far as possible, to eat everything he kills" over the winter, a time when it had previously taken few lives, it would not grow even warier than it already was. But this did not save Corbett from press criticism, and while he was away it struck ten more times.[22]

Hearing on his return that it was again on the east side of the Alaknanda, Corbett once more ordered the two bridges to be closed at night. He then wasted much time and effort following up false alarms for, as he put it, "in an area in which an established man-eater is operating everyone suspects their own shadows, and every sound heard at night is attributed to the man-eater."[23]

For a long time, too, fortune continued to favor the leopard. One night, sitting over a cow it had killed, Corbett was just taking aim at the man-eater when a woman knocked at a nearby house, scaring it away.[24]

Another time Corbett, Ibbotson, and one of Ibbotson's men tethered a goat on a hillside near a village where the man-eater had been heard calling, then hid behind some rocks. The goat bleated loudly until sunset, then fell silent and stared into the bushes; the leopard had arrived. But so cautious and well camouflaged was it that even with a telescopic sight they could not spot it.

With the light fading fast they untied the goat and hurried back toward the village, but on the way the goat ran off. Foolishly they went after it; and found it lying twitching on a path, blood oozing from its throat. It was as if, Corbett recalled, the leopard was saying: "Here, if you want your goat so badly, take it; and as it is now dark and you have a long way to go, we will see which of you lives to reach the village."[25]

They scrambled down the hillside, Corbett striking match after match to light the way until, hearing their yells, some villagers came to meet them carrying lanterns and flaming torches. Only the matches had saved them from attack, reckoned Corbett. The next morning paw prints confirmed that the man-eater had followed them all the way.

The same morning came news that it had killed a man at dawn outside his hut four miles up the hill. It had carried him to a nearby hollow where it had eaten his throat, jaw, and part of one shoulder and thigh before being disturbed by the wailing of the man's family. Corbett and Ibbotson laced the corpse with cyanide. The next day they found the man-eater had eaten from the other shoulder and leg, carefully avoiding all the poisoned parts.

Undeterred, they laced the remains with yet more cyanide, burying it deep in all the uneaten flesh. Corbett then scattered dry leaves on the path into the hollow and hid in a tree overlooking the path. At eight the bark of a kakar broke the silence: the man-eater was back. At ten it barked again. The leopard had been feeding for two hours, long enough to have poisoned itself many times over. Four hours later Corbett heard the crunch of dry leaves as it staggered along the path. But it stopped while still out of sight and for several more hours he waited motionless, rifle at the ready. At last it moved again, only to veer away, drink thirstily from a spring, and slope off into the night.

Inspection of the remains at dawn confirmed that the leopard had eaten plenty of poisoned flesh. Corbett had 200 men beat the area, and traced the man-eater to the narrow opening of a cave, where it had vomited up its victim's undigested toes before squeezing inside. Corbett sealed the opening with wire mesh, and for ten days returned morning and evening with the growing expectation that the stench of decay would tell him the leopard had died inside. But on the tenth day came news that it had killed again, five miles away.

There must have been another opening further up the hill, reasoned Corbett, but it was no wonder, he wrote, that the people of Garhwal credited the leopard with supernatural powers and clung to the belief that nothing but fire would rid them of this evil spirit.[26]

8. Beast Master

Corbett and Ibbotson found the latest victim, an old woman, under a hillside rose bush; stripped of all clothing, partially eaten, and strewn with white petals. The sight hardened their resolve: this time, surely, the killer would pay. They buried the gin-trap on the side they expected the leopard to approach from, and covered it with twigs and leaves. They lashed two rifles to saplings, aiming them where Corbett judged that the leopard would lie down to resume its meal if it missed the gin-trap, and ran a length of new silk fishing line from the corpse to the triggers. Then for good measure they laced the remains with cyanide.

As they stood back to inspect their handiwork they saw that the man-eater might just approach from the opposite side, so they blocked off that route with thorn bushes. Finally they made a straw-cushioned machan in the nearest suitable tree, 200 yards away, and surrounded it with wire mesh in case the leopard turned its attentions on *them*.

Sunset saw them lying on their stomachs, rifles at the ready. Darkness fell with no sign of the leopard. They put away their guns. One chance had gone. It began to rain, and Corbett fretted about the weight of water on the plate setting off the trap, and about the new line shrinking and prematurely triggering the rifles. But then they heard angry roars: the man-eater was in the trap.

As they rushed over, the roars stopped, and Corbett feared the leopard had escaped. Alas, he was right; the sprung trap lay empty. Examination of the scene showed how "cunning," or lucky, the man-eater had been. According to Corbett, after scattering the thorn bushes it pulled the corpse towards the rifles, slackening the lines, and ate all around the poisoned flesh. And then, when caught in the trap, the jaws closed over its leg just where one tooth was missing, only grazing the leg, and allowing it to tear itself free.

Far from making Corbett lose heart, this latest failure strengthened his resolve. But one thing he determined. With Ibbotson again returning to his other duties he would forget all about traps and poison, and put all his faith in his own trusty aim.[27]

For a long time, though, fortune continued to favor the leopard. One night, Corbett sat up a hillside tree over a goat, which instead of bleating plaintively curled up and went to sleep.

Presumably this was because the goat knew Corbett was there. Usually, wrote Best, sportsmen settled in position before their helpers tethered a goat. The helpers then made sure it saw them leave. Only this way would it think it had been abandoned, and so begin to bleat. To achieve the same end, he added, some sportsmen used a she-goat with kids, and tied the latter up out of her sight, but within her hearing. Forsyth noted that when shikaris in India used a goat or dog as bait, they tugged on a line to a fish hook thrust through the animal's ear to make it bleat or yelp. Similarly, when Malays baited tiger traps with a dog or goat, to make it yelp or bleat they shoved a sharp stone into its ear, reported Shuttleworth, or, according to Locke, tied it up with its back feet clear of the ground.[28]

Anyway, with the goat fast asleep, Corbett tried calling up the man-eater by imitating the cries of a leopardess on heat. To his delight the leopard called back, from around 400 yards away. A few minutes later it called again from around 200 yards away. Corbett called in turn, and the leopard answered from only 100 yards away. By this time he had his thumb on the switch of the flashlight lashed to his rifle. But when the man-eater called from just sixty yards away, a real leopardess further up the hill answered, and it headed off in her direction. Another chance had gone begging.[29]

Despite many such setbacks, Corbett never gave up hope. Whenever his failures got him down, the faith most Garhwalis still had in him restored his spirits. The man-eater's time had not come yet, they would say, but when it did, the sahib would triumph.[30]

But by April 1926, Corbett was exhausted, and Ibbotson was under increasing pressure to open the field to other sportsmen, should any actually wish to enter it. Corbett urged his old friend to give him ten more days, during which he would sit up every night at the village of Golabrai on the pilgrim road, for he knew from studying its tracks that the leopard passed that way every five days or so. Reluctantly Ibbotson agreed, and the two men set out to build a machan in a mango tree overlooking the road there.

By this time Corbett was friends with a pandit — a Hindu scholar — who owned a shelter in Golabrai, and in vain the man pleaded with him not to risk his life in the tree, for like the old Banjara he doubted even Corbett could match this particular evil spirit.

The pandit's fears were well founded. One evening five years earlier a woman sought refuge in his shelter, fearing she might not reach home before dark. Seeing how scared she was he showed her to the back of the room. Relieved, she fell asleep, with no fewer than fifty pilgrims between herself and the door. That night the leopard crept in, stepped over row after row of sleeping pilgrims, seized her by the throat, retraced its steps, and carried her off into the night; all without waking a soul.

Then, one evening a short while after that, ten weary pilgrims who had traveled all the way from Madras (now Chennai) knocked on his door. Fearing for his reputation if the man-eater struck again in the shelter, he let them share his own room. It was a hot, stuffy night, and eventually he risked stepping out onto the veranda for some fresh air. (According to Corbett, that is: sometime in the 1990s the pandit's grandson, Tilak Ram Deoly, told Prabir Nath Banerji from Meerut that his grandfather was recklessly sleeping on the veranda.[31])

No sooner was he outside than the man-eater leapt up and buried its teeth in his throat. Clinging on to the roof supports he managed to bring his bare feet up underneath its body, and with one mighty adrenalin-fueled kick sent it flying down the steps. As he swooned against the railing it sprung again and sank its claws into his left arm, raking him from elbow to wrist as it fell back under its own weight. But before it could pounce for a third time the pilgrims managed to drag him inside and bolt the door. They spent the rest of the night screaming as the growling leopard clawed repeatedly at the door, while the pandit lay bleeding and gasping for breath through the gaping holes in his throat. Six months later he returned home from hospital in Rudraprayag with his health broken and his hair turned gray; one of only two people known to have survived being attacked by the man-eater. The first thing he did was to board up his veranda.

Now the good sahib was intent on exposing himself to the demon for ten consecutive nights, and nothing he said could dissuade him from his task.

But the ten nights passed uneventfully, with no sign of the leopard.

Ibbotson and Corbett discussed the situation. No other sportsmen had come forward and they were both reluctant to concede defeat. Yet Ibbotson was neglecting his other duties, and Corbett needed to visit his farm in Africa. In two minds as to what to do, they agreed to postpone their decision until morning. At that, Corbett said he would try for the leopard one last time.

When Ibbotson accompanied his friend to the mango tree that evening they came across two snakes fighting in a field, and Ibbotson clubbed one of them to death as it tried to escape across the road. It was a good omen.

After Ibbotson had left, the pandit told Corbett that no fewer than 150 exhausted pilgrims were insisting on spending the night in his shelter, despite his urging them not to. Corbett told him to warn them to stay together and on no account go outside after dark.

Not heeding this sound advice, just after nine o'clock that night, May 1, 1926, a pilgrim left the shelter carrying a lantern and crossed the road to answer a call of nature. A few minutes later, to Corbett's immense relief, he returned: but even as he went back in, dogs started barking nearby. The man-eater had arrived. Corbett could see nothing in the dark, but from the changing direction of the racket he knew the leopard was coming down the road; and when the noise stopped he knew it was crouching out of sight somewhere very close by. There followed long minutes of silence and uncertainty. Was it stalking the shelter or the goat? Corbett shut his eyes to focus on his hearing; his rifle, flashlight attached, pointing toward the goat. Then came a rush, and the tinkling of the goat's bell. Flicking on the flashlight, he glimpsed the leopard and fired. As he did so the flashlight gave out, and the man-eater disappeared into the night.

At the sound of the shot a wave of murmuring came from the shelter, and the pandit called out from the doorway. But Corbett was listening intently for the leopard and did not answer. Fearing the worst, the pandit hastily fastened his door, and the murmuring died down. It was ten o'clock. Corbett thought he had heard a faint gurgle coming from the field across the road, but there was nothing he could do now until dawn. He made himself as comfortable as he could on his machan, lit a cigarette — the first of many, no doubt — and waited.

At daybreak Corbett climbed stiff-limbed down to the ground, where he was greeted by a bleat from the unharmed goat; and, joy of joys, by an inch-wide streak of blood at the side of the road. Following the trail he found the leopard only fifty yards away, slumped in a hollow. The most hated and feared animal in India was dead at last (Figure 14). But here was no fiend, reflected the sportsman: "Here was only an old leopard ... whose only crime — not against the laws of nature, but against the laws of man — was that he had shed human blood, with no object of terrorizing man, but only in order that he might live; and who now, with his chin resting on the rim of the hole and his eyes half-closed, was peacefully sleeping his long last sleep."[32]

Just then the pandit arrived. On seeing the "evil spirit" lying dead on the ground he put his hands together and bent to bow his forehead to Corbett's feet, but the sportsman gently restrained him. Moments later Corbett's men appeared, and within minutes the leopard was being borne triumphantly towards Rudraprayag and the Inspection Bungalow, with the goat trotting happily at Corbett's heels.

When Ibbotson woke to the sound of knocking on the door he real-

Figure 14. The celebrated "White Sadhu" of Kumaon, Jim Corbett, with the infamous Rudraprayag leopard, said to be a transformed sadhu. The man-eater claimed an officially recorded 125 victims, but in all likelihood, thought Corbett, it killed many more, no remains of whom were ever found, and who therefore could not officially be recorded as victims (© Oxford University Press, India).

ized straight away what must have happened. Leaping out of bed he embraced his old friend and danced wildly around the dead leopard on the veranda.

After Corbett had enjoyed a well-earned cup of tea and hot bath, thousands of people came from miles around to see the dead fiend and lay flowers at the sportsman's feet. "I have on other occasions witnessed gratitude," he recalled, "but never as I witnessed it that day at Rudraprayag."[33]

When he gave the goat back to its previous owner, the man proudly put a brass collar on it: and earned a tidy sum from exhibiting it for the rest of its days.[34]

Corbett was already revered in Kumaon, but his triumph at Rudraprayag earned him legendary status there, and after more success, against man-eating tigers, in 1944 his best-selling *Man-Eaters of Kumaon* brought him worldwide fame. But when India gained independence in 1947, he and Maggie felt they had no choice but to leave the beloved country of their births. Corbett no longer owned the Tanganyika farm so they settled in Kenya, near relatives and friends. All his years in Africa Corbett dreamt of seeing Kumaon again, but ill health and a foreboding about the changes he would find there prevented him from ever returning. He died in Kenya, and was buried there, in 1955.

Corbett finished his book about the Rudraprayag leopard in his last year in India. If in it he occasionally embroidered the truth, the fact remains that for many years the people of Rudraprayag celebrated his triumph with an annual fair, and even now he is revered as the White Sadhu throughout Kumaon. In his introduction to the 1990 edition of Corbett's 1953 book *Jungle Lore*, Booth recalled visiting Kaladhungi, where Corbett had his winter home, when filming a drama-documentary of the sportsman's life five years earlier. Booth was sheltering from the fierce summer sun when a frail old man arrived and asked to meet "Carpet Sahib," as Kumaonis have always called Corbett. Booth naturally thought he wanted to meet the actor playing Corbett, Frederick Treves, who readily agreed to see him. The old man staggered over with his walking stick, salaamed, and made to press his forehead to Treves's feet: he truly believed Corbett had come back to Kumaon, as long foretold. It transpired he had walked sixty miles in two days to reach Kaladhungi.

Prabir Nath Banerji found Golabrai much the same in the 1990s as it was in 1926. The pandit's house — complete with veranda — and his pilgrim shelter both still stood. So did the mango tree from which Corbett fired the fatal shot. It still bore fruit.[35]

Chapter Notes

Preface

1. Endicott 1979, 53.
2. Bernatzik 1958, 139.
3. Hazewinkel 1963, 45.
4. Wessing 1995.

Introduction

1. Baudesson 1919, 132–3; Thurston 1912, 76; Powell 1957, 203. Strictly speaking, a shikari is anyone who hunts, but the term was most often used to distinguish a native hunter from a colonial sportsman.
2. Christopher 1916, 103.
3. Keyser 1922, 118. Keyser also wrote about this anonymously in *Asia* magazine: see British Official 1922, 472.
4. Hicks 1910, 625.
5. See, for example, Campbell 1842, 198–9; Shakespear 1860, 77; Baker 1887, 113; Brown 1887, 198; Webber 1902, 27. One exception was sportsman George Sanderson, who as early as the 1870s wrote that there appeared to be no foundation for the belief: Sanderson 1879, 279. Folklorist William Crooke wrote in 1906 that the belief had been quite abandoned among colonials by then, but it was one J. B.H. Thurston, a rubber-planter in Malaya, held as late as 1939, and renowned India born and bred sportsman Kenneth Anderson as late as 1959: Crooke 1906, 478; Thurston 1939, 244; Anderson 1959, 18. The notion that tigers suck blood is sometimes stated as fact even today: see for example, Boomgaard 2001, 27.
6. Locke 1954, 123, 154–5 (and see also Wessing 1986, 85–6, and British Official 1922, 471); Burton 1931, 29–30.

Chapter 1

1. For more on the history of British policy in the CPs, see McEldowney, 1980.
2. Forsyth 1871, 31, 98.
3. Northern, Aryan Indians once called the whole region Gondwana, or "land of the Gonds." Gondwana in turn is the source of the name Gondwanaland, the enormous continental area—comprising Africa, Antarctica, South America, Arabia, and Australia, as well as the Indian peninsula—that geologists reckon existed in the southern hemisphere following the break up of the supercontinent Pangaea in the Mesozoic era.
4. Forsyth 1871, 98, 364.
5. Forsyth 1871, 158–9.
6. Sleeman 1844, vol. 1, 56.
7. Elwin 1936, 47, 130.
8. Elwin 1936, 230; McMillan 1906, preface.

Notes. Chapter 1

9. Forsyth 1871, 35, 37; Forsyth 1879, 61–2. See also Louden 1889, 98–101.
10. Forsyth 1871, 314.
11. Forsyth 1871, 41, 386; Forsyth 1879, 59–60.
12. Forsyth 1871, 17–18, 34, 37.
13. Forsyth 1871, 3, 36–7, 310.
14. Forsyth 1871, 95, 312. Today Pachmarhi is a tourist resort. A vantage point there, Priyadarshini Point, was formerly called Forsyth Point. Bison Lodge is now a forestry museum.
15. Forsyth 1871, 95, 165–76.
16. Forsyth 1871, 75, 101–3, 111–20; Elwin 1936, 22.
17. Forsyth 1871, 120–1.
18. Hicks 1910, 5, 663–4.
19. Hicks 1910, 1–3.
20. Hicks 1910, 2, 6–7, 310, 602–5.
21. Hicks 1910, 5, 223–4, 236–7; Forsyth 1871, 285.
22. Hicks 1910, 5, 404.
23. Hicks 1910, 152.
24. Forsyth 1871, 40, 293–4.
25. See, for example, Forsyth 1871, 210–1, 308; Forsyth 1879, 64–5; Sanderson 1879, 182–4, 210; and Louden 1889, 100.
26. Hicks 1910, 601.
27. Hicks 1910, 511.
28. Forsyth 1871, 212–3, 309; Forsyth 1879, 68–9.
29. Hicks 1910, 613.
30. Hicks 1910, 8–9.
31. Hicks 1910, 573–83.
32. Hicks 1910, 293–5; Forsyth 1871, 281–2. See also Shakespear 1860, 122. For more about "the Lalla," see Sterndale 1877, 447–50.
33. The situation in remote areas of Java and Sumatra was much the same: see Boomgaard 2001, 44–58.
34. Best 1931a, 24. Mail runners are still employed in remote areas of the mountainous state of Himachal Pradesh in northern India. See http://maddy06.blogspot.com/2010/01/dak-harkaka.html.
35. Hicks 1910, 5–6; Webber 1902, 27–8: Fayrer 1875, 41.
36. Best 1931a, 23; Best 1931b, 200–1.
37. Best 1931a, 21, 22. See also, for example, Brown 1887, 185, and Durand 1911, 118.
38. Brown 1887, 161; Forsyth 1871, 290.
39. Hicks 1910, 254–9.
40. For more on the causes of man-eating, see: Forsyth 1871, 295–6; Sanderson 1879, 270–2; Hicks 1910, 619–32; Perry 1964, 185–205; Bright 2000, 30–64, 95–7.
41. Best 1935, 92–4.
42. Baker 1887, 205; Durand 1911, 67–8.
43. Hicks 1910, 568; Boomgaard 2001, 61; Carrington Turner 1959, 94.
44. See, for example: Shakespear 1860, 77; Gordon Cumming 1871, 6; Forsyth 1871, 264; Sterndale 1877, 321; Sanderson 1879, 41, 360; Mervyn Smith 1904, 103; Hicks 1910, 120.
45. Hicks 1910, 120; Best 1931a, 21, 22; Burton 1931, 106; Baker 1887, 206.
46. McNair 1878, 82.
47. Keyser 1922, 116–7. Keyser also wrote about this anonymously in *Asia* magazine: see British Official 1922, 471–2. The Dutch likewise offered rewards for tigers in Java and Sumatra: see Boomgaard 2001, 63, 87–106.
48. Thurston 1939, 215.
49. Fayrer 1875, 54; Forsyth 1871, 254, 305; Sanderson 1879, 267, 269.
50. Aflalo 1912, 13.
51. Eardley-Wilmot 1910, 89.

Chapter 2

1. Baudesson 1919, 41–2.
2. Wessing 1995.
3. Wessing 1986, 11; Wessing 1995; Bakels 1993.
4. Jacobs 1894, vol. 1, 297.
5. Skeat 1900, 158–9.
6. Hutton 1921b, 317–8.
7. Hutton 1921a, 261–2. See also Mao 2009.
8. Locke 1954, 167–8. Wessing heard a similar story in Semarang in Central Java: Wessing 1986, 21–2.
9. Wessing 1986, 10.
10. Butler 1854, 152.
11. Boomgaard, 2001, 59–60, 167–70, 185; Locke 1954, 92.
12. Juliusson 1974, 77
13. Latham 1859, 341–3; Sterndale 1877, 433–4.
14. Bhagvat 1972, 69.
15. Laufer 1912, 182–3.
16. Russell and Hira Lal 1916, vol. 3, 510.
17. Hurgronje 1906, vol. 2, 301; Wessing 1986, 43–4.
18. Wessing 1986, 64–5; Bradley 1929, 115.
19. Locke 1954, 163.
20. Locke 1954, 163–5.
21. Fuchs 1960, 395.
22. Mitra 1894b, 162–3.
23. Russell and Hira Lal 1916, vol. 3, 112–3.
24. Forsyth 1871, 141; Mitra 1894a, 47–8.
25. Elwin 1936, 91–2. See also Fuchs 1960, 156.
26. Crooke 1894, 36, 72; Crooke 1907, 231.
27. Mirat 1894a, 55. For more about Daskin Ray and Bonobibi, see Montgomery 1995, 105–19, 142–62.
28. Baudesson 1919, 129, 133, 199.
29. Perry 1964, 196–7.
30. Crooke 1894, 74–7, 322; Mitra 1894a, 48. In other versions of the Dulha Deo legend, Dulha was killed by lightning or turned to stone.
31. Wessing 1986, 29–30.
32. Best 1931b, 170; Hanley 1928, 34–40. See also Bernatzik 1958, 157.
33. Fuchs 1960, 540.
34. Wessing 1986, 46, 48–9.
35. Kipling 1891, 355; Wessing 1986, 42–3; Boomgaard 2001, 181.
36. Hurgronje 1903, 240–1.
37. Wessing 1986, 35, 40, 87–8.
38. Endicott 1979, 95, 113, 136.
39. Wessing 1986, 42.
40. Wessing 1986, 67–70.
41. Endicott 1981, 16.
42. Bakels 1993. See also Endicott 1979, 135–8.
43. Endicott 1981, 154–67.
44. Wavell 1958, 103–9.
45. Wessing 1986, 39–40, 47.
46. Wessing 1986, 50; Boomgaard 2001, 42; Endicott 1979, 59; Skeat 1900, 163; Locke 1954, 170.
47. Hicks 1910, 489–92.
48. Hicks 1910, 128–36. Russell and Hira Lal 1916, vol. 2, 388.
49. Sanderson 1879, 306–13.
50. Russell and Hira Lal 1916, vol. 3, 130; Young 1962, 18, 75; Carey 1976, 97; Hutton 1921a, 345; Wessing 1986, 47, 90–1; Baze 1957, 61; Bakels 1993.

51. Bernatzik 1958, 148; Endicott 1979, 79.
52. Thurston 1912, 75; Bakels 1993; Christopher 1916, 171–2; Endicott 1979, 80.
53. Crooke 1894, 36; Crooke 1907, 231; Bakels 1993; Endicott 1979, 79.
54. Mitra 1894b, 158; Locke 1954, 162.
55. Baker 1887, 110–12.
56. Chu 2009, 128; Sanderson 1879, 297 (and see also 317). See also Crooke 1894, 321; Hicks 1910, 126; Durand 1911, 88; Gouldsbury 1913, 123.
57. Tod 1873, vol. 2, 566; Boomgaard 2001, 172–3.
58. Wessing 1995.
59. Banner 1927, 154–5.
60. Boomgaard 2001, 44, 56; Raffles 1835, vol. 1, 356–7; Bradley 1929, 120.
61. Hurwood 1968, 36–9.
62. Wessing 1986, 17–18.
63. Bakels 1993.
64. Wessing 1986, 21.
65. Skeat 1900, 167–70.
66. Flower 1920, 894; Wessing 1986, 20.
67. Hutton 1921a, 92, 159, 182, 262, 340; Hutton 1921b, 200.
68. Lehman 1963, 183.
69. Crooke 1906, 483. For more on the Narimangala, see Ganapathy 1967, 87–91.
70. Elwin 1936, 57.
71. Bernatzik 1958, 213; Flower 1920, 894; Campbell 1885, 281; Caldwell 1925, 48.
72. de Groot 1901, 179; Crooke 1894, 322; Thurston 1912, 76 (and see also Mitra 1894a, 47); Hutton 1921b, 165.
73. Banner 1927, 159; Mitra 1894a, 60; Hicks 1910, 115; Burton 1931, 94; Sanderson 1879, 313.
74. Burton 1931, 94; Crooke 1894, 323; Mitra 1894a, 56; Locke 1954, 176–7.
75. Christopher 1916, 87–8; Mitra 1894a, 58; Mitra 1984b, 161; Russell and Hira Lal 1916, vol. 3, 564; Wessing 1986, 52; Baudesson 1919, 130–1. See also Campbell 1842, 405; Baker 1887, 140; Brown 1887, 165–7; British Official 1922, 472.
76. Christopher 1916, 87–8; Hicks 1910, 115; Burton 1931, 94; Powell 1957, 220; Crooke 1894, 324; Baudesson 1919, 132; Durand 1911, 107.
77. Brown 1887, 166; Coleman 1832, 321.
78. Crooke 1894, 324; Sanderson 1879, 313; Forsyth 1871, 451; Elliott 1973, 53.
79. Thurston 1912, 78–9; Mitra 1912, 373; Skeat 1900, 159. Skeat did not know what kind of tree los came from, but it may have been screw pine, also known as pandan (*Pandanus* species): see Wessing 1986, 12.
80. Locke 1954, 155; Skeat and Blagden 1906, vol. 2, 294; Skeat 1900, 167–8.
81. Baudesson 1919, 200.
82. Marsden 1784, 253; Skeat 1900, 157–8; Skeat 1901, 26–7; Locke 1954, 158–9. See also Wessing 1986, 97–8, and Boomgaard 2001, 190–1.
83. Boomgaard 2001, 186, 191.
84. Skeat 1901, 26–7; Wessing 1986, 98.
85. Endicott 1979, 43; Mills 1926, 247–8 (see also Changkija 2007); Wessing 1986, 46.
86. Wessing 1986, 69; Bakels 1993.
87. Wessing 1986, 66–7.

Chapter 3

1. See for example, Berger, S., "Villagers in Terror of Victims' Ghosts," *Daily Telegraph*, January 15, 2005. http://www.telegraph.co.uk/news/worldnews/africaandindianocean/1481268/Villagers-in-terror-of-victims-ghosts.html
2. Elwin 1936, 27–8.
3. Webber 1902, 27–8; Hicks 1910, 239–52.
4. Baker 1887, 92–3, 130; Shakespear 1860, 78. See also Hicks 1910, 126, and Durand 1911, 87, 144.

5. Crooke 1894, 321; Crooke 1907, 238–9.
6. Russell and Hira Lal 1916, vol. 3, 81, 196.
7. Wessing 1986, 89–90; Baudesson 1919, 47; Baze 1957, 53.
8. Many colonials in India believed man-eaters preferred the flesh of sweet-smelling vegetarian Hindus to that of rank, meat-eating Europeans: see, for example, Best 1931b, 186.
9. Baudesson 1919, 42–51.
10. Locke 1954, 159; de Groot 1907, 554–63.
11. de Groot 1907, 557–8. In a variant of this tale the sprite, grief-stricken at the tiger's death, enters its mouth and turns into a piece of jade: de Groot 1907, 563. Another story tells how a farmer losing geese to a tiger outwitted the animal's protective ch'ang kwei by decoying the spirit with fruit: de Groot 1907, 558–9.
12. de Groot 1907, 556–7.
13. Hamel 1915, 11; Perry 1964, 221; Anderson 1959, 157.
14. Perry 1964, 214.
15. Anderson 1967, 36.
16. Shakespear 1860, 66–7.
17. Locke 1954, 125, 133–53.
18. Christopher 1916, 71, 104; Fayrer 1875, 41–2. In Java it was traditional to place an old horse at the rear of a column in case a tiger attacked: Boomgaard 2001, 44.
19. Handley 1933, 154–5.
20. Russell and Hira Lal 1916, vol. 3, 564; Fuchs 1960, 356; Thurston 1939, 244.
21. Taylor 1956, 77; Bradley 1929, 114.
22. See, for example, Hicks 1910, 151.
23. King Martin 1935, 163–6.
24. Baze 1957, 61.
25. Best 1931a, 24.
26. Wessing 1986, 15–16.
27. Forsyth 1871, 338–42.
28. Burton 1931, 137; Powell 1957, 217.
29. Hanley 1961, 94–8.
30. Eardley-Wilmot 1910, 51–3.
31. Campbell 1885, 281; Allámi 1891, vol. 2, 224; Burton 1931, 135; Elwin 1936, 93.
32. Crooke 1907, 256; Crooke 1894, 168.
33. Russell and Hira Lal 1916, vol. 3, 560.
34. Russell and Hira Lal 1916, vol. 3, 112–3; Fuchs 1960, 156, 355–67.
35. Russell and Hira Lal 1916, vol. 2, 84–5; Russell and Hira Lal 1916, vol. 3, 95; Forsyth 1871, 363; Elwin 1936, 28.
36. Sanderson 1879, 280; Perry 1964, 144. See also Durand 1911, 48–9.
37. McMillan 1906, 36–9; Forsyth 1871, 363.
38. McMillan 1906, 37.
39. Russell and Hira Lal 1916, vol. 2, 274, 536.
40. Gribble 1944, 4–5.

Chapter 4

1. Durand 1911, 108.
2. Crawford 1909, 64–77. When the 17-year-old Shivaji, with his mother's encouragement, vowed in 1645 to free his country from Mogul tyranny, he did so in a temple of Shiva and sealed the vow by spilling blood from his thumb onto the linga there.
3. Wessing 1986, 28.
4. Burton 1931, 136.
5. Hanley 1928, 8–15.
6. Crooke 1894, 168–71. In Malay tradition, women who die in childbirth become vampires called *pontianaks*.
7. Wessing 1986, 30; Anderson 1959, 87.
8. Best 1931b, 66–82.

9. Forsyth reckoned muknas were braver than the female elephants usually used in tiger beats; Forsyth 1871, 287. Maharajahs used males with tusks, for prestige, but as famed elephant catcher Sanderson noted, sportsmen generally used females because they were easier to procure. While a well-trained male was much braver than a female, wrote Sanderson, an ill-disciplined male was a much greater risk, being likely to attack the tiger and shake you out of the howdah: Sanderson 1879, 90.
10. Baudesson 1919, 133.
11. Forsyth 1871, 291–305.
12. Hicks 1910, 530–42.

Chapter 5

1. Wessing 1986, 55–7.
2. Endicott 1979, 132.
3. de Groot 1907, 547–8.
4. Locke 1954, 155; Wessing 1986, 25–6. Because of the symptoms manifested, like convulsions and foaming at the mouth, rabies has often been proposed to explain cases of lycanthropy, but Adam Douglas argues persuasively that the conditions are only superficially similar: Douglas 1993, 235–6.
5. Burton 1931, 195.
6. Wessing 1986, 69, 94, 102.
7. Hutton 1921b, 200–8; Hutton 1921a, 243–4.
8. Carey 1976, 106–9.
9. Carey 1976, 112; Wessing 1986, 115–6.
10. *Transformations* 1989, 26.
11. Thurston 1912, 260.
12. Wessing 1986, 85.
13. Wessing 1986, 80–1, 85. Bradley heard similar tell of grave-robbing tjindaks; Bradley 1929, 120.
14. Crooke 1896, vol. 4, 405; Elwin 1936, 230.
15. Lindskog 1954, 147.
16. Jacobs 1894, vol. 1, 292–3.
17. de Groot 1898, 555; de Groot 1901, 166; Allen 1989; Crooke 1894, 168–9; Knebel 1899, 576–7; Hazeu 1899, 690.
18. Schilling 1957, 66–76, 170–86.
19. de Groot 1907, 559–60. See also Ashley 2001, 150. According to Ashley, this story was first told by eighth-century scholar Tai Fu, in his *Kuang-i chi*, or *Great Book of Marvels*.
20. Skeat 1900, 161; Wessing 1986, 82–3. See also Knebel 1899, 579.
21. Winter 1902, 85; Boomgaard 2001, 195–6; Knebel 1899, 579; Hazeu 1899, 690; Banner 1927, 153.
22. Wessing 1986, 80; Pliny the Elder, *The Natural History*, viii, 22 (see also Douglas 1993, 122); Hurwood 1968, 34; Hutton 1921a, 243–4.
23. Douglas 1993, 43; Sullivan 1981, 73.
24. Winter 1902, 85; Knebel 1899, 579; Wessing 1986, 69.
25. Hazeu 1899, 692; Wessing 1986, 56.
26. Wessing 1986, 69 (and see 38); de Groot 1901, 164–5, 179–80.
27. Masters and Houston 1966, 58–60.
28. As quoted by Jeremy Lennard in the *Guardian*, 25 January 1999: http://www.guardian.co.uk/world/1999/jan/25/jeremylennard1.
29. Bakels 1993.
30. Wessing 1986, 76, 78.
31. Sullivan 1981, 73.
32. Douglas 1993, 38; Knight 1995.
33. For an overview, see http://faculty.washington.edu/chudler/moon.html.
34. Hutton 1921b, 202.
35. Packer *et al* 2011.

36. de Groot 1907, 548.
37. Sullivan 1981, 74.
38. Hutton 1921b, 202–5.
39. Hutton 1921b, 203.
40. Fürer-Haimendorf 1946, 209; Endicott 1979, 133, 140.
41. Hutton 1921b, 203, 204.
42. Hutton 1921b, 201; Hammond 1991, 92–3 (for Fan Tuan story); de Groot 1901, 180.
43. de Groot 1898, 558; de Groot 1901, 166–7; Douglas 1993, 147–8; Burton 1931, 195–6; Wessing 1986, 84 (and see 94).
44. bin Ahmad 1922; Wessing 1986, 67, 86; Endicott 1979, 126, 132, 137–8.
45. Bradley 1929, 115–6.
46. Bradley 1929, 119–20; Douglas 1993, 255–6; Crooke 1894, 194; Locke 1954, 174–5.
47. Douglas 1993, 128, 131, 147–9.
48. Boomgaard 2001, 205.
49. Bradley 1929, 115, 120; Banner 1927, 153–4.
50. de Groot 1907, 545–6; de Groot 1901, 167–9. A "peck" is a measure equivalent to eight quarts.
51. Bernatzik 1958, 101, 106. See also Mitra 1894b, 161.
52. de Sélincourt 1972, 306; Hutton 1921a, 243; Bernatzik 1958, 159.
53. Wilken 1884, 21–2; van Balen 1914, 358; Hazewinkel 1963, 45.
54. Hazeu 1899, 689–90.
55. Knebel 1899, 573–7; Rouffaer 1899, 69; van Balen 1914, 357. Boomgaard notes that there never was in fact a village called Gadungan in Lodoyo, and says the real Gadungan was some 50 km away to the northwest of Mount Kelud, just southeast of Paré, but had disappeared from maps by 1938: Boomgaard 2001, 200. But that particular Gadungan *does* appear to be marked on modern maps, along with one just to the south of Mount Kelud, and another just to its west, plus one elsewhere in Java, and another in Bali.
56. Winter 1902, 85; Knebel 1899, 581.
57. Boomgaard 2001, 205–6.
58. de Groot 1907, 546.
59. de Groot 1907, 553–4.
60. de Groot 1907, 551–2.
61. de Groot 1901, 180–1.
62. Aylesworth 1970, 62–3.
63. Bradley 1929, 117–9.
64. Bradley 1929, 116.
65. de Groot 1901, 168.
66. Kingscote and Sastri 1890, 119–30; Wessing 1986, 86–7. Georgiana wrote under the name Mrs. Howard Kingscote.
67. van Ossenbruggen 1916, 188.

Chapter 6

1. For more about Clifford, see Barr 1977.
2. Allen 1983, 44. See also Barr 1977, 147.
3. Barr 1977, 149.
4. Clifford 1897, 63; Locke 1954, 154.
5. Clifford 1897, preface, 196–209. For a slightly revised version of this story, see Clifford 1916, 148–65.
6. Perry 1964, 204.
7. Hicks 1910, 548–52.
8. Endicott 1979, 140–1.
9. Skeat and Blagden 1906, vol. 2, 191, 227–9.
10. Boomgaard 2001, 194–5; Bradley 1929, 120; Skeat 1900, 160–1.
11. Wessing 1986, 29, 101; Wessing 1992, 301. See also Boomgaard 2001, 145–66.
12. Boomgaard 2001, 194–5, 205–6.

13. Keyser 1922, 119–20. Keyser also wrote about this anonymously in *Asia* magazine: see British Official 1922, 472.
14. For more about Swettenham, see Barr 1977.
15. Swettenham 1895, introduction.
16. Swettenham 1900, preface; Ainsworth 1933, 90–1.
17. Swettenham 1895, 200–1.
18. Hooker 1855, vol. 7, 45 (for Motley); van Hasselt 1882, 75.
19. Boomgaard 2001, 192; Wessing 1986, 96.
20. van Hasselt 1882, 75–6.
21. Bakels 1993; Marsden 1784, 253; Presgrave 1821, 41.
22. Swettenham 1895, 200–1.
23. Clifford 1897, 62–77. For a slightly revised version of this story, see Clifford 1916, 40–55.
24. Locke 1954, 157–8.
25. Clifford 1897, 65; Clifford 1916, 41 (for Sayong detail).
26. Maxwell 1907, 265–79.
27. Lumsden Milne 1932, 84–6.
28. Burke and Burke 1935, 39; Thurston 1939, 245.
29. Shuttleworth 1965, 86; Alexander 1935, 56.
30. Alexander 1935, 56–7.
31. Clifford 1897, 65.
32. Maxwell 1907, 269–71.
33. Shuttleworth 1965, 86.
34. Swettenham 1895, 200; Clifford 1897, 63; Maxwell 1907, 269; Rathborne 1898, 105; Bradley 1929, 120; Thurston 1939, 245; Newbold 1839, vol. 2, 192; Wessing 1986, 102. The Burkes thought Kerinci weretigers were "men who leave their bodies to take the form of tigers in order to satisfy their lust for human flesh and blood." Burke and Burke 1935, 39.
35. Wavell 1958, 87–92.
36. Harrison 1969, 13–22.
37. Skeat 1900, 159–60. Locke heard much the same story in Terengganu; Locke 1954, 161–2.

Chapter 7

1. Best 1931b, 157–70.
2. Inglis 1878, 63–4.
3. Inglis 1878, 64–6.
4. Fuchs 1960, 531–40; Juliusson 1974, 77.
5. Ball 1880, 115.
6. Oman 1908, 306; Crooke 1894, 357; Fuchs 1960, 533–4.
7. Chandola 2007, 160–72.
8. Ball 1880, 115.
9. *The Asiatic Annual Register…for the Year 1801*, vol. 3, 91–2.
10. Dalton 1872, 199.
11. Ball 1880, 115–6.
12. Oman 1908, 307. See also Crooke 1894, 367–8, and Ball 1880, 115.
13. Fuchs 1960, 534, 538.
14. Malcolm 1823, vol. 2, 213.
15. Fuchs 1960, 538–40.
16. Dalton 1872, 200–1.
17. Thurston 1912, 261–2.
18. Campbell 1885, 257–8.
19. Hanley 1961, 53–66.
20. Crooke 1894, 363–4.
21. Baigas likewise tipped their arrowheads with aconite: see Forsyth 1871, 360. A staple ingredient of the psychoactive potions and ointments traditionally used by European witches, Indian aconite—*Aconitum ferox*—is notoriously the world's most poisonous plant.

22. Mervyn Smith 1904, 101-11.
23. For a critical examination of O'Donnell's depiction of women in *Werewolves*, see Bourgault du Coudray 2006, 48-9.
24. Thurston 1912, 199-207.
25. Thurston 1912, 260.
26. O'Donnell 1972, 20-9.
27. Sterndale 1877, 368-84, 452; Sterndale 1884, 180-1. The title alone of Sterndale's 1877 work is thought to have inspired Rudyard Kipling, who never actually visited Seoni, to set *The Jungle Book* (1894) there.
28. Forsyth 1871, 321-2.
29. See, for example, Campbell 1842, 411-2; Forsyth 1871, 318-9; Sanderson 1879, 327-32; Baker 1887, 192-4; Hicks 1910, 178-84.
30. Sterndale 1884, 178-84. Shakespear sensibly distinguished only between the "leopard" (the cheetah) and the "panther" (the leopard); Shakespear 1860, 101.
31. Sterndale 1877, 447.
32. de Groot 1901, 168.
33. Hicks 1910, 127.
34. Elwin 1936, 174-6.
35. Skeat and Blagden 1906, vol. 2, 191.
36. Anderson 1967, 121-5.
37. McEldowney 1980; Baker 1887, 141.
38. For an examination of the often contradictory images of the Pathan character painted by the British, see Lindholm 1980.
39. Sterndale 1877, 368-84, 452; Sterndale 1884, 180-1.
40. Kipling 1891, 358; Hamel 1915, 11; Candee 1927, 63-4.
41. Wessing 1986, 101; Boomgaard 2001, 199.
42. Sleeman 1844, vol. 1, 165-7.
43. Sleeman 1844, vol. 1, 162-5.
44. Jepson 1936, 37-40.
45. Jepson 1936, 40.
46. Thurston and Rangachari 1909, 342, 354, 357, 359.
47. Thurston and Rangachari 1909, 353-4. See also Anantha Krishna Iyer, L. K. 1909, vol. 1, 167, and Thurston 1912, 260-1.
48. Powell 1957, 217-8.
49. Russell and Hira Lal 1916, vol. 2, 129-34.
50. Best 1931b, 16-19.
51. Russell and Hira Lal 1916, vol. 3, 153, 247-8.
52. Russell and Hira Lal 1916, vol. 2, 14.
53. Elwin 1936, 185.
54. Burton 1931, 134.
55. Sanderson 1879, 44-7.
56. Mitra 1894a, 49.
57. Forsyth 1871, 386.
58. Russell and Hira Lal 1916, vol. 2, 537; Kipling 1891, 355-6; Mitra 1894a, 54-5.
59. Eardley-Wilmot 1910, 53; Kipling 1891, 356.

Chapter 8

1. Corbett 1948, 7-12, 125.
2. Corbett 1948, 4-5; Best 1935, 253. Best recalled the dry bed of the Narmada that winter being littered with so many thousands of rotting corpses, some partially eaten by turtles and crocodiles, that he could not help treading on them when out shooting snipe.
3. Corbett 1948, 18-22.
4. India born and bred sahibs like Corbett and Anderson were once known as Anglo-Indians, but confusingly that term later came to be applied to people of mixed British and Indian parentage.

Notes. Chapter 8

5. Booth 1986, 101; Kala 1999, 43.
6. Burke and Burke 1935, 25. For more on: the Champawat tigress, see Corbett 1946, 1–27; the Muktesar tigress, see Corbett 1954, 43–68; the Panar leopard, see Corbett 1954, 69–92.
7. Corbett 1954, 69, 77. Burke and Burke 1935, 25–6.
8. Corbett 1954, 70.
9. The Burkes reckoned it took to man-eating after being wounded in one paw, but Corbett, who mentioned no injuries, reckoned it got the taste for human flesh from eating corpses during a cholera outbreak: Burke and Burke 1935, 25; Corbett 1946, xvii (Author's Note).
10. Booth 1986, 128; Corbett 1948, 40; Corbett 1946, 60.
11. Corbett 1946, 163.
12. Corbett 1948, 22–3; Booth 1986, 133–4, 190.
13. Corbett 1948, 40.
14. Corbett 1948, 44.
15. Corbett 1948, 46.
16. Corbett 1948, 34–68.
17. Corbett 1948, 13–17.
18. Corbett 1948, 18; Crooke 1896, vol. 2, 60.
19. Corbett 1948, 28–30.
20. Corbett 1948, 68. Banner and Locke respectively noted that in Java and the Malay Peninsula it was bad luck if a snake crossed your path from left to right, but good luck if it crossed from right to left: Banner 1927, 159; Locke 1954, 186.
21. Corbett 1948, 18, 69.
22. Champion 1934, 113–4; Corbett 1948, 69, 70.
23. Corbett 1948, 71.
24. Corbett 1948, 72.
25. Corbett 1948, 83.
26. Corbett 1948, 90.
27. Corbett 1948, 80–101.
28. Best 1931a, 66; Forsyth 1871, 324; Shuttleworth 1965, 36; Locke 1954, 36.
29. Corbett 1948, 113–5, 120–1.
30. Corbett 1948, 106, 121.
31. Banerji.
32. Corbett 1948, 151–2.
33. Corbett 1948, 155.
34. Corbett 1948, 141–55.
35. Banerji.

References

Aflalo, Frederick G. 1912. *A Book of the Wilderness and Jungle*. London: S. W. Partridge.
Ainsworth, Leopold. 1933. *The Confessions of a Planter in Malaya*. London: H. F. & G. Witherby.
Alexander, Patrick. 1935. "Spirits of the Malay Jungles." *Asia* 35.1: 54–7.
Allámi, Abul F. 1891. *Aín I Akbari*. Translated by Henry S. Jarrett. Calcutta: Asiatic Society of Bengal.
Allen, Benedict. 1989. *Hunting the Gugu*. London: MacMillan.
Allen, Charles, ed. 1983. *Tales from the South China Seas*. London: André Deutsch and the BBC.
Anantha Krishna Iyer, L. K. 1909. *The Cochin Tribes and Castes*. Madras: Higginbotham.
Anderson, Kenneth. 1959. *The Black Panther of Sivanipalli*. London: George Allen & Unwin.
———. 1967. *The Tiger Roars*. London: George Allen & Unwin.
Ashley, Leonard R. N. 2001. *The Complete Book of Werewolves*. New York: Barricade Books.
The Asiatic Annual Register ... for the Year 1801. 1802. London: J. Debrett and T. Cadell Jun. & W. Davies.
Aylesworth, Thomas G. 1970. *Werewolves and Other Monsters*. Reading, MA: Addison-Wesley.
Bakels, Jet. 1993. "The Tiger by the Tail: On Tigers, Ancestors and Nature Spirits in Kerinci." *Tiger Beat* (Tiger Species Survival Plan newsletter) 6, 2, 21–3. Minnesota Zoo.
Baker, Edward B. 1887. *Sport in Bengal*. London: Ledger, Smith.
Balen, J. Hendrik van. 1914. *De Dierenwereld van Insulinde in Woord en Beeld*, vol. 1. Deventer: Van der Burgh.
Ball, Valentine. 1880. *Jungle Life in India*. London: Thomas de la Rue.
Banerji, Prabir N. "In the Footsteps of Corbett." http://web.archive.org/web/20011008181616/http://www.topwritecorner.com/essays/banerjiessay.html
Banner, Hubert S. 1927. *Romantic Java*. London: Seeley, Service.
Barr, Pat. 1977. *Taming the Jungle*. London: Secker & Warburg.
Baudesson, Henri. 1919. *Indo-China and Its Primitive People*. Translated by E. Appleby Holt. London: Hutchinson.
Baze, William. 1957. *Tiger! Tiger!* Translated by H. M. Burton. London: Elek Books.
Bernatzik, Hugo A. 1958. *The Spirits of the Yellow Leaves*. Translated by E. W. Dickes. London: Robert Hale.
Best, James W. 1931a. *Indian Shikar Notes*. Lahore: Pioneer Press. First published in 1920.
———. 1931b. *Tiger Days*. London: John Murray.
———. 1935. *Forest Life in India*. London: John Murray.
Bhagvat, Durga. 1972. "Folk Tales of Central India." *Asian Folklore Studies* 31.2: 1–89.
bin Ahmad, Zainul-Abidin. 1922. "The Tiger Breed Families." *Journal of the Straits Branch of the Royal Asiatic Society* 85: 36–9.
Boomgaard, Peter. 2001. *Frontiers of Fear*. New Haven, CT: Yale University Press.
Booth, Martin. 1986. *Carpet Sahib*. London: Constable.
———. 1990. Introduction to *Jungle Lore*, by Jim Corbett. India: Oxford University Press.
Bourgault du Coudray, Chantal. 2006. *The Curse of the Werewolf*. London, New York: I. B. Tauris.
Bradley, Mary H. 1929. *Trailing the Tiger*. New York, London: D. Appleton.

References

Bright, Michael. 2000. *Man-Eaters*. London: Robson Books.
British Official. 1922. "In Tiger-Haunted 'Kampongs.'" *Asia* 22.6: 467–72, 496.
Brown, James M. 1887. *Shikar Sketches*. London: Hurst & Blackett.
Burke, Redmond St. G., and Norah Burke. 1935. *Jungle Days*. London: Stanley Paul.
Burton, Reginald G. 1931. *A Book of Man-Eaters*. London: Hutchinson.
Butler, John. 1854. *Travels and Adventures in the Province of Assam*. London: Smith, Elder
Caldwell, Harry R. 1925. *Blue Tiger*. London: Duckworth. First published in 1924.
Campbell, James M. 1885. *Notes on the Spirit Basis of Belief and Custom*. Bombay: Government Central Press.
Campbell, Walter. 1842. *The Old Forest Ranger*. London: How and Parsons.
Candee, Helen C. 1927. *New Journeys in Old Asia*. New York: Frederick A. Stokes.
Carey, Iskander. 1976. *Orang Asli — The Aboriginal Tribes of Peninsular Malaysia*. Kuala Lumpur: Oxford University Press.
Carrington Turner, Joshua E. 1959. *Man-Eaters and Memories*. London: Robert Hale.
Champion, Frederick W. 1934. *With a Camera in Tiger-Land*. London: Chatto & Windus. First published in 1927.
Chandola, Sudha. 2007. *Entranced by the Goddess*. Loughborough, Leicestershire, England: Heart of Albion Press.
Changkija, Narola. 2007. "From Oral Tale to Graphic Novel: Re-animating the Tiger-Soul." Ph.D. dissertation, Griffith University. http://www4.gu.edu.au:8080/adt-root/public/adt-QGU20090218.072040/index.html.
Christopher, Sydney A. 1916. *Big Game Shooting in Lower Burma*. Rangoon: Burma Pictorial Press.
Chu, Man-ping. July 2009. "Chinese Cultural Taboos That Affect Their Language and Behavior Choice." *Asian Culture and History* 1.2: 122–139.
Clifford, Hugh. 1897. *In Court and Kampong*. London: Grant Richards.
_____. 1916. *The Further Side of Silence*. London: Curtis Brown.
Coleman, Charles. 1832. *The Mythology of the Hindus*. London: Parbury, Allen.
Corbett, Jim. 1946. *Man-Eaters of Kumaon*. London: Oxford University Press. First published in 1944, in India.
_____. 1948. *The Man-Eating Leopard of Rudraprayag*. Madras: Oxford University Press.
_____. 1953. *Jungle Lore*. London: Oxford University Press.
_____. 1954. *The Temple Tiger*. London: Oxford University Press.
Crawford, Arthur. 1909. *Legends of the Konkan*. Allahabad: Pioneer Press.
Crooke, William. 1894. *An Introduction to the Popular Religion and Folklore of Northern India*. Allahabad: Government Press.
_____. 1896. *The Tribes and Castes of the North-Western Provinces and Oudh*. Calcutta: Office of the Superintendent of Government Printing.
_____. 1906. *Things Indian*. London: John Murray.
_____. 1907. *Natives of Northern India*. London: Archibald Constable.
Dalton, Edward T. 1872. *Descriptive Ethnology of Bengal*. Calcutta: Office of the Superintendent of Government Printing.
de Groot, Jan J. M. 1898. "De weertijger in onze koloniën en op het Oostaziatische vasteland." *Bijdragen tot de Taal-, Land- en Volkenkunde van Nederlandsch-Indië* 49: 549–85.
_____. 1901. *The Religious System of China*, vol. 4, book 2. Leiden: E. J. Brill.
_____. 1907. *The Religious System of China*, vol. 5, book 2. Leiden: E. J. Brill.
de Sélincourt, Aubrey, trans. 1972. *Herodotus: The Histories*. London: Penguin Classics.
Douglas, Adam. 1993. *The Beast Within*. London: Orion (Chapmans). First published in 1992.
Durand, Edward. 1911. *Rifle, Rod, and Spear in the East*. London: John Murray.
Eardley-Wilmot, Sainthill. 1910. *Forest Life and Sport in India*. London: Edward Arnold.
Elliott, James G. 1973. *Field Sports in India*. London: Gentry Books.
Elwin, Verrier. 1936. *Leaves from the Jungle*. London: John Murray.
Endicott, Kirk M. 1979. *Batek Negrito Religion*. Oxford: Clarendon Press.
_____. 1981. *An Analysis of Malay Magic*. Kuala Lumpur: Oxford University Press. First published in 1970.
Fayrer, Joseph. 1875. *The Royal Tiger of Bengal*. London: J. & A. Churchill.
Flower, Henry C. 1920. "On the Trail of the Lord Tiger." *Asia* 20.9: 893–897.

References

Forsyth, James. 1863. *The Sporting Rifle and Its Projectiles*. London: Smith, Elder.
_____. 1871. *The Highlands of Central India*. London: Chapman & Hall.
Forsyth, James (Rev.). 1879. *Memoir of the Late Captain James Forsyth*. Pages 59–70 of History of the Berwickshire Naturalist's Club, vol. 8, 1876–1878. Alnwick: Henry Hunter Blair, 1879.
Fuchs, Stephen. 1960. *The Gond and Bhumia of Eastern Mandla*. Bombay: Asia Publishing House.
Fürer-Haimendorf, Christoph von. 1946. *The Naked Nagas*. Calcutta: Thacker, Spink.
Ganapathy, Bacamada D. 1967. *Kodavas (Coorgs)*. Mangalore: Sharada Press.
Glasfurd, Alexander I. R. 1903. *Leaves from an Indian Jungle*. Bombay: Times Press, "Times of India" Office.
Gordon Cumming, William. 1871. *Wild Men and Wild Beasts*. Edinburgh: Edmonston & Douglas.
Gouldsbury, Charles E. 1913. *Tigerland*. London: Chapman & Hall.
Gribble, R. H. 1944. *Out of the Burma Night*. Calcutta: Thacker, Spink.
Hamel, Frank. 1915. *Human Animals*. London: William Rider & Son.
Hammond, Charles E. 1991. "An Excursion in Tiger Lore." *Asia Major* 4.1: 87–100.
Handley, Leonard M. H. 1933. *Hunter's Moon*. London: MacMillan.
Hanley, Maurice P. 1928. *Tales and Songs from an Assam Tea Garden*. Calcutta: Thacker, Spink.
Hanley, Patrick. 1961. *Tiger Trails in Assam*. London: Robert Hale.
Harrison, Horace L. H. 1969. *The Sarong and the Kris*. Lymington, Hampshire, England: Nautical Publishing in association with George C. Harrap, London.
Hasselt, Arend L. van. 1882. *Volksbeschrijving van Midden-Sumatra*. Leiden: E. J. Brill.
Hazeu, Godard A. J. 1899. "Eenige mensch-dierverhalen uit Java." *Bijdragen tot de Taal-, Land- en Volkenkunde van Nederlandsch-Indië* 50: 688–94.
Hazewinkel, J. C. 1963. *De Tijger in het Volksgeloof*. Whittier, CA: American Tong-Tong.
Hicks, Frederick C. 1910. *Forty Years Among the Wild Animals of India*. Allahabad: Pioneer Press.
Hooker, W. J., ed. 1855. *Hooker's Journal of Botany and Kew Garden Miscellany*. London: Lovell Reeve.
Hurgronje, Dr. Christiaan S. 1903. *Het Gajoland en Zijne Bewoners*. Batavia: Landsdrukkerij.
_____. 1906. *The Achehnese*. Translated by A. W. S. O'Sullivan. Leiden: E. J. Brill.
Hurwood, Bernhardt J. 1968. *Vampires, Werewolves, and Ghouls*. New York: Ace Books.
Hutton, John H. 1921a. *The Angami Nagas*. London: MacMillan.
_____. 1921b. *The Sema Nagas*. London: MacMillan.
Inglis, James ("Maori"). 1878. *Sport and Work on the Nepaul Frontier*. London: MacMillan.
Jacobs, Dr. Julius, 1894. *Het Familie-en Kamponleven op Groot-Atjeh*. Leiden: E. J. Brill.
Jepson, Stanley, ed. 1936. *Big Game Encounters*. London: H. F. & G. Witherby.
Juliusson, Per. 1974. "The Gonds and Their Religion." Ph.D. dissertation, University of Stockholm.
Kala, Durga C. 1999. *Jim Corbett of Kumaon*. New Delhi: Ravi Dayal. First published in 1979.
Keyser, Arthur L. 1893. *An Adopted Wife*. London: Griffith Farran.
_____. 1922. *People and Places*. London: John Murray.
King, Martin, D. 1935. *The Ways of Man and Beast in India*. London: Wright & Brown.
Kingscote, H., and Pandit Natesa Sastri. 1890. *Tales of the Sun*. London: W. H. Allen.
Kipling, John L. 1891. *Beast and Man in India*. London: MacMillan.
Kipling, Rudyard. 1894. *The Jungle Book*. New York: MacMillan.
Knebel, Josef. 1899. "De weertijger op Midden-Java, den Javaan naverteld." *Bijdragen tot de Taal-, Land- en Volkenkunde van Nederlandsch-Indië* 41: 568–87.
Knight, Chris. 1995. *Blood Relations*. New Haven, CT: Yale University Press.
Latham, Robert G. 1859. *Ethnology of India*. London: John van Voorst.
Laufer, Berthold. 1912. *Jade: A Study in Chinese Archaeology and Religion*. Chicago: Field Museum of Natural History Publication 154, Anthropological Series vol. 10.
Lehman, Frederic K. 1963. *The Structure of Chin Society*. Urbana: University of Illinois Press.
Lindholm, Charles. 1980. "Images of the Pathan: The Usefulness of Colonial Ethnography." *European Journal of Sociology* 21: 350–61.
Lindskog, Birger. 1954. "African Leopard Men." Ph.D. dissertation, University of Uppsala.
Locke, Arthur. 1954. *The Tigers of Trengganu*. London: Museum Press.
Louden, David. 1889. *The History of Morham*. Haddington: W. M. Sinclair.

References

Lumsden Milne, Betty. 1932. *Damit and Other Stories.* Singapore: Rickard.
Malcolm, John. 1823. *A Memoir of Central India.* London: Kingsbury, Parbury & Allen.
Mao, X. P. 2009. "The Origin of Tiger, Spirit and Humankind: A Mao Naga Myth." *Indian Folklife* 33: 10.
Marsden, William. 1784. *The History of Sumatra.* London: private printing.
Mason, Philip. 1989. *The Wild Sweet Witch.* New Delhi: Penguin Books India. First published, under the name Philip Woodruff, in 1947 (London: Jonathan Cape).
Masters, Robert E. L., and Jean Houston. 1966. *The Varieties of Psychedelic Experience.* New York: Dell Publishing.
Maxwell, George. 1907. *In Malay Forests.* Edinburgh: William Blackwood & Sons.
McEldowney, Philip F. 1980. "Colonial Administration and Social Developments in Middle India: The Central Provinces, 1861–1921." Ph.D. dissertation, University of Virginia. http://www.lib.virginia.edu/area-studies/SouthAsia/Ideas/CP/intro.html.
McMillan, Archibald W. 1906. *Jungle Pioneering in Gondland.* London: Morgan & Scott.
McNair, Frederick. 1878. *Perak and the Malays.* London: Tinsley Brothers.
Mervyn Smith, A. 1904. *Sport and Adventure in the Indian Jungle.* London: Hurst & Blackett.
Mills, James P. 1926. *The Ao Nagas.* London: MacMillan.
Mitra, Sarat Chandra. 1894a. "Indian Folk-Beliefs About the Tiger." *The Journal of the Anthropological Society of Bombay* 3.1: 45–60.
———. 1894b. "Additional Folk-Beliefs About the Tiger." *The Journal of the Anthropological Society of Bombay* 3.3: 158–63.
———. 1912. "The Tiger in Malay Folklore, Proverbial Philosophy, and Folk-Medicine." *The Journal of the Anthropological Society of Bombay* 9.6: 369–377.
Montgomery, Sy. 1995. *Spell of the Tiger.* New York: Houghton Mifflin.
Newbold, Thomas J. 1839. *Political and Statistical Account of the British Settlements in the Straits of Malacca.* London: John Murray.
O'Donnell, Elliott. 1934. *Strange Cults and Secret Societies of Modern London.* London: Philip Allan.
———. 1972 (revised edition). *Werwolves.* New York: Wholesale Book Corp. First published in 1912.
Oman, John C. 1908. *Cults, Customs and Superstitions of India.* London: T. Fisher Unwin.
Ossenbruggen, Frederik D. E. van. 1916. "Het priemitieve denken zoals dit zich uit voornamelijk in pokkengebruiken op Java en Elders." *Bijdragen tot de Taal-, Land- en Volkenkunde van Nederlandsch-Indiö* 71: 1–370.
Packer, Craig, Alexandra Swanson, Dennis Ikanda, and Hadas Kushnir. 2011. "Fear of Darkness, the Full Moon and the Nocturnal Ecology of African Lions." *PLoS ONE* 6(7): e22285. http://www.plosone.org/article/info%3Adoi%2F10.1371%2Fjournal.pone.0022285
Perry, Richard. 1964. *The World of the Tiger.* London: Cassell.
Pliny the Elder. *The Natural History of Pliny.* Edited by John Bostock and Henry T. Riley. London: H. G. Bohn, 1855–57.
Powell, Arthur N. W. 1957. *Call of the Tiger.* London: Robert Hale.
Presgrave, Edward. 1821. *Account of a Journey from Manna to Passumah Lebar, and the Ascent of Gunung Dempo, in the Interior of Sumatra.* Fort Marlborough: Mission Press.
Raffles, Sophia, ed. 1835. *Memoir of the Life and Public Services of Sir Thomas Stamford Raffles.* London: James Duncan. First published (in one volume) in 1830.
Rathborne, Ambrose B. 1898. *Camping and Tramping in Malaya.* London: Swan Sonnenschein.
Rouffaer, Gerret P. 1899. "Matjan gadoengan." *Bijdragen tot de Taal-, Land- en Volkenkunde van Nederlandsch-Indiö* 50: 67–75.
Russell, Robert V., and Rai Bahadur Hira Lal. 1916. *The Tribes and Castes of the Central Provinces of India.* London: MacMillan.
Sanderson, George P. 1879. *Thirteen Years Among the Wild Beasts of India.* London: W. H. Allen. First published in 1878.
Schilling, Ton. 1957. *Tigermen of Anai.* Translated by E. W. Dickes. London: George Allen & Unwin. First published as *Tijgermensen van Anai,* Amsterdam: H. Meusenhoff, 1952.
Shakespear, Henry. 1860. *The Wild Sports of India.* Boston: Ticknor and Fields.
Shuttleworth, Charles. 1965. *Malayan Safari.* London: Phoenix House.
Skeat, Walter W. 1900. *Malay Magic.* London: MacMillan.

References

_____. 1901. *Fables & Folk-Tales from an Eastern Forest*. Cambridge: Cambridge University Press.

_____, and Charles O. Blagden. 1906. *Pagan Races of the Malay Peninsula*. London: MacMillan.

Sleeman, William H. 1844. *Rambles and Recollections of an Indian Official*. London: J. Hatchard & Son.

Sterndale, Robert A. 1877. *Seonee*. London: Samson Low, Marston, Searle, & Rivington.

_____. 1884. *Natural History of the Mammalia of India and Ceylon*. Calcutta: Thacker, Spink.

Sullivan, J. P., trans. 1981. *Petronius, the Satyricon and Seneca, the Apocolocyntosis*. London: Penguin Classics.

Swettenham, Frank A. 1895. *Malay Sketches*. London: John Lane.

_____. 1900. *The Real Malay*. London: John Lane.

Taylor, Mary L. 1956. *The Tiger's Claw*. London: Burke.

Thurston, Edgar. 1912. *Omens and Superstitions of Southern India*. London: T. Fisher Unwin.

Thurston, Edgar, and Rangachari, K. 1909. *Castes and Tribes of Southern India*. Madras: Government Press.

Thurston, J. B. H. (as told to Lieut. H. S. Mazet). 1939. "Tiger! Tiger! Adventures with Man-Eaters in the Malay Jungles." *Natural History* 44.4: 213–6, 244–5.

Tod, James. 1873. *Annals and Antiquities of Rajasthan*. Madras: Higginbotham.

Transformations. 1989. Alexandria, VA: Time-Life Books.

Wavell, Stewart. 1958. *The Lost World of the East*. London: Souvenir Press.

Webber, Thomas W. 1902. *The Forests of Upper India*. London: Edward Arnold.

Wessing, Robert. 1986. *The Soul of Ambiguity: The Tiger in Southeast Asia*. (Special Report no.24, Northern Illinois University monograph series on Southeast Asia.) Wyoming: Cellar Bookshop.

_____. 1992. "A Tiger in the Heart: The Javanese Rampok Macan." *Bijdragen tot de Taal-, Land- en Volkenkunde van Nederlandsch-Indiö* 148.2: 287–308.

_____. 1995. "The Last Tiger in East Java: Symbolic Continuity in Ecological Change." *Asian Folklore Studies* 54: 191–218.

Wilken, George A. 1884. *Het animisme bij de volken van den Indischen Archipel*. Amsterdam: de Bussy.

Winter, Johannes W. 1902. "Beknopte beschrijving van het hof Soerakarta in 1824." *Bijdragen tot de Taal-, Land- en Volkenkunde van Nederlandsch-Indiö* 54: 15–172.

Young, Gordon. 1962. *The Hill Tribes of Northern Thailand*. Bangkok: Journal of the Siam Society Monograph 1.

Zangwill, Israel. 1921. *The Voice of Jerusalem*. New York: MacMillan.

Glossary

adat	custom (Malay world)
ashram	religious retreat (India)
bagh	tiger/big cat (India)
bagh-mari	professional tiger-hunter (India)
bhut	spirit or ghost (India)
burning-ghat	riverside cremation site (India)
ch'ang kwei	spirit or ghost of someone killed by tiger (China)
charpoy	light bedstead (India)
chowkidar	night-watchman (India)
CPs	Central Provinces (India)
datok	chief (Malay world)
dukun	Kerinci shaman (Sumatra)
hookah	tobacco-pipe with water to cool smoke (India)
howdah	seat on elephant's back for two or more riders (India)
Huzoor	Lord (India)
kampong	village (Malay world)
macan gadongan	weretiger (Java)
machan	platform hide (India)
mahout	elephant-driver (India)
mahua spirit	alcoholic drink made from Mahua tree flowers (India)
moonshee	secretary (India)
nullah	dry river-bed, ravine (India)
olthwa	bloodsucking tiger-form creature (central India)
pawang	shaman (Malay world)
penghulu	headman (Malay world)
peon	orderly (India)
sadhu	Hindu ascetic (India)
shadow soul	type or aspect of soul that can wander free from body
shaitan	demon (India)
shikar	hunting (India)
shikari	native hunter (India) — but see first *Introduction* note, page 185
sirih	quid made from betel leaf etc. (Malay world)
sola topi	pith helmet
spiritual shapeshift	control of real animal by shadow soul
thakur	chief (India)
tjindak	weretiger (Sumatra)
Tuan	Lord (Malay world)
UPs	Upper Provinces (India)

Index

Numbers in ***bold italics*** indicate pages with illustrations.

Aceh 27–8, 32, 36–7, 38, 40, 80, 93, 94, 97, 98
Actaeon 97, 100, 158
adat 40, 121
An Adopted Wife 6
Aflalo, Frederick 25
Afzal Khan 154
Ageng Tirtayasa 32
Aghoris 164
Ahiri Forests 15–16
Aín I Akbari 67
Ainsworth, Leopold 119
Alexander, Patrick 127–8
Allah 30, 32, 54, 126, 129, 134
Allámi, Abul Fazl 67
Allen, Benedict 98
Almora 170–1, 176
Amarkantak 9, 10, 61, 164, 165
Anai Ravine 98
Anantha Krishna Iyer, L. K. 162
Anderson, Kenneth 59, 81, 154–5, 170
Angami Nagas 29, 47–8, 100, 108
Anhui 101
Antaeus 100
Ao Nagas 54
Artemis 97, 100, 103
Arthur, Sir George 77
Arthur's Seat ***77***
Asia [magazine] ***50***, 127
The Asiatic Annual Register 138
Assam 28, 35, 116, 141–4
Aylesworth, Thomas 109–10

Badarinath 167, 176
bagh nakh **154**
Bagheshwar 33–4, 35, 69
Baigas 10, 11, 31, 34, 35, 36, 61, 67–74, 81–3, 97, 135–6, 137, 158, 162, 163
Bakels, Jet 28, 38, 42, 43, 46–7, 54, 102, 120–1

Baker, Edward 22, 43–4, 56, 156
Balaghat 11, 19, 63, 72–3, 74, 162
Bali 1
Balinese tiger 1
Ball, Valentine 137, 138, 139
Banda Aceh 40, 80
Banerji, Prabir Nath 181, 184
Bangalore 59, 81
"Banije-balingka" 120
Banjaras 19, 66, 81–3, 86, 175, 180
Banner, Hubert 44–5, 50, 69, 99, 107
Bansapti Ma 34, 43
Banten 32, 40
banyans 162–3, 164
Bastar 33
Batak 47, 56, 63
Batek 1, 37–8, 40, 42, 43, 54, 93, 104, 106, 116–17
Batu Balok 54, 93
Baudesson, Henri 6, 27, 34–5, 51, 53, 56–8, 61, 85
Baze, William 42, 56, 57, 63
Beast and Man in India 93
begar 10–11
beggars 107, 108, 118, 165
Benares 157
Bencoolen 45, 53; *see also* Bengkulu
Bengal 23, 50, 51, 56, 164
Bengal & North-Western Railway 169
Bengkulu 45, 120
Bentong 124–6, 127, 129
Berar 35, 95, 105
berhantu 38–40
Bernatzik, Hugo 1, 42, 48, 107, 108
Besar 40
Best, James 19, 20, 21–2, 23, 35, 63, 81–3, 135–6, 137, 163, 168, 179–80
Betul 21, 84–7, ***88***
Bhagavathi 162
Bhagvat, Durga 31

Index

Bhairava 13, 66, 164, 167
Bhandara 156
Bhatras 75
Bhawani 33, 66, 79, 80, 88, 92, 103
Bhils 10, 13, 31, 67
Bhoksas 66, 176
Bhumias *see* Baigas
Bhutan 3
bhuts 34, 55–6, 80–1
bidog 96
Bihar 33, 138–9, 169
Bilaspur 20, 81–3, 135–6, 153, 163
bin Ahmad, Zainul-Abidin 105–6
Bison Lodge 12–14, 15
black leopards 81, 105
black tigers 81–3
b'ladau 154
Blagden, Charles 52–3, 117, 154
Blaine, Mahlon **4**, *123*
Blandas 52
Blitar 108
Bloomfield, Alfred 11
boarmen 120
bobongkongs 36, 54
bomohs 38
Bonobibi 34
Boomgaard, Peter 22, 29, 30, 37, 40, 44, 45, 53, 99, 107, 108, 117, 118, 120
Booth, Martin 169, 184
Borpatra 65
bounties 22–3
Bradley, Mary Hastings 32, 45, 62, 106, 107, 110, 117, 129
Brahma 164
Brahmans 30, 80, 111–12, 141; *see also* priests (Brahman)
Brooke-Wavell, Stewart *see* Wavell, Stewart
Broun, Heywood C. 55
Brown, James 21, 51
Bukit Nanas 132
Bukit Tinggi *see* Fort de Kock
Burke, Norah 127, 170
Burke, Redmond 127, 170
Burma 1, 6, 43, 48, 50–1, 60–1, 75
Burton, Reginald 7, 23, 50, 51, 65, 67, 80, 94–5, 105, 164
"bush soul" 38; *see also* "shadow soul"
Butler, Capt. John 29

Caldwell, Harry 49
Cambodia 1, 109–10, 158
Cambridge Expedition 52, 117
Campbell, Sir James MacNabb 48, 67, 141
Candee, Helen Churchill 158
Carey, Iskander 42, 95–6
Carrington Turner, Joshua 22

Caspian tiger 1
Castes and Tribes of Southern India 162
Central Java 38, 108
Central Provinces Gazette 63
Champawat 170
Champion, Frederick 169
Chamunda 164
Chanda 15, 16, 19, 62
Chandola, Sudha 138
ch'ang kweis 58–9, 99
Chathan 162
chedipes 141, 147, 153
"Chenaku" 119–20
Chenchus 59
Chennai 181
Cheros 69
Chhindwara 41
Chins 48
Chota Nagpur 61, 72, 139, 144–7
Chowgarh 171
Christopher, Sydney 6, 43, 51, 61
Chu, Man-ping 44
churels 80–1, 98
"Cindaku" 120
Clifford, Hugh 4, 113–16, 119, 121–4, 128, 129, 132
clothing 99, 100, 108–9, 128–9, 153
Cochin 162
Coimbatore 154–5
Coleman, Charles 51
Cooper's Hill 15, 19
Coorg *see* Kodagu
Copeland, Barbara 129–30, 132
Corbett, Jim 169–84 (**183**)
Corbett National Park 170
Crawford, Arthur 76–80
crickets 36
Crooke, William 34, 35, 43, 48, 49, 50, 51, 56, 68–9, 81, 97, 98, 106, 138, 144, 176
crooked feet 98, 107

Dalton, Col. Edward 139, 140–1, 147
Damoh 59
Danteshwari 33
Dasara 33
Daskin Ray 34, 35
Daya 38, 54, 81, 95, 97, 100, 101
Deccan 67
de Groot, Jan 49, 58–9, 65, 93–4, 98, 99, 101, 103, 105, 107, 108–9, 110, 124, 153
Devi 138
Devi Patan 66
dhobis 160–2
Dhuma 150
diet doubling 104–5, 117, 121, 127–8, 145, 175
Dinajpur 23

202

Index

Dindori 20
Dioscorea 158
Diwali 139–40
Douglas, Adam 100, 102, 105, 106
Duc-Thay 34–5, 56, 85
Dugger Forests 116
dukuns 38, 102, 117, 120–1
Dulha Deo 35
Dumals 75
Durand, Edward 22, 51, 76
Durga 33, 103, 139–40, 148, 162, 164
Duri Gond clan 33

Eardley-Wilmot, Sainthill 25–6, 66–7, 166
Earth Mother 33, 103, 138, 139, 140, 148; see also Mother Earth
East Java 108
Eastern Ghats 149
Eldorado Syndrome 32, 107
Elliott, Charles 69, 72
Elliott, Maj.-Gen. James 52
Elwin, Verrier 10, 13, 34, 48, 55, 67–8, 72, 97, 154, 164
Endicott, Kirk 1, 37, 38, 40, 42, 43, 54, 93, 104, 106, 116–17
Eriligarus 67

fakirs 107, 165, 166
familiars 36, 38–40
fatalism 62–3
Fayrer, Joseph 20, 25, 61, 135
Firebird 99
Flower, Henry C. 47, 50, 57
"follow tigers" 32–3, 54
Forsyth, Capt. James 10, 11–14, 17, 18–19, 21, 25, 33, 51–2, 63–4, 72, 73, 75, 76, 84–7, **88**, 151, 152, 157, 165, 180
Forsyth, Rev. James 11, 17
"Forsyth's shell" 14
Fort de Kock 62, 106, 110
Fuchs, Father Stephen 33, 36, 61, 69–75, 106, 137, 138, 139, 140
The Further Side of Silence **4, 123**

Gadungan 108
Ganges 9, 15, 138, 157, 168, 172, 173
Ganjam 42
Gansam Deo 35
Garhwal 167–84
Garnier, Gilles 107
Garos 51
Gayo 27–8, 37, 42, 46, 51
gaze aversion 98, 107
Gir forest 161
Glasfurd, Alexander 167
Godavari 141, 147
Golabrai 180–4

gold 106, 117–18
Gonda 66
Gonds 10, 12–13, 30–1, 33–4, 35, 36, 41, 42, 48, 51, 55, 56, 61, 62, 64, 67, 69–74, 80–1, 84–92, 97, 116, 136, 137, 138, 145, 151, 152–3, 156, 157, 159, 162
Gobar 40, 54
Gopeng 132
Great Book of Marvels 105
Great Indian Peninsular Railway 11
Great Van 35
Great White Tiger 80
Grenier, Jean 105, 107
Gribble, Capt. R.H. 75
Guangdong 108
Gugu 98
Gunong Ledang 53
Gwalior 161

Hailey National Park 170
haj 118
Halbas 56
Hamel, Frank 59, 157–8
Handley, Leonard 9, 61
Hanley, Maurice 35, 65–6, 80–1, 141–4, 170
Hanley, Patrick *see* Hanley, Maurice
hantu belian 38, 53–4; *see also* tiger-form ancestors
Harrison, Horace 132–4
Hawkins, Gerald 130–2
Hazeu, Godard 98, 99, 100, 108
Hazewinkel, J.C. 2
Hecate 97
Herodotus 107
Hicks, Frederick 7, 14–18, 21, 22, 41, 50, 51, 56, 75, 87–92, 113, 116, 153, 164
The Highlands of Central India 17, 76, **88**
Hira Lal, Rai Bahadur 31, 33, 41, 42, 51, 56, 61, 69, 72, 75, 163, 164, 165
Hos 49
Hoshangabad 12, 15, 56, 69
Houston, Jean 101
Hubback, Theodore 170
Hubei 93–4, 98, 107
Hunan 110, 124
Hunter's Moon 9
Hurgronje, Dr. Christiaan Snouck 32, 37
Hurwood, Bernhardt J. 45–6, 100
Hutton, John 28–9, 42, 47–8, 49, 100, 103, 104, 105, 108

Ibbotson, Bill 169, 173–84
Ibok 106
Ikanda, Dennis 103
The Illustrated London News **49**
ilmu 38
In Court and Kampong 113

Index

incense 38, 117, 162; *see also* joss-sticks
Indian Mutiny 9, 11, 60, 152
Indo-Chinese tiger 170
Inglis, James 136–7
iron 44, 69, 71–4, 111–12
Iskandar Muda 80

Jabalpur 11–12, 13, 21, 59, 63–4, 158
Jabalpur Samachar 156
jackals 44, 48, 52, 56
Jacobs, Julius 28, 97–8
Jambi 28, 119
Jarays 48
Javan tiger 1
Jepara 108
Jepson, Stanley 161
Jerangau 60
jewelry 106, 145
Jhansi 160–1
Jiangxi 107, 153
Jijabai 76–80
Johor 53, 113
joss-sticks 39; *see also* incense
Juangs 49
Jugra 47, 118
Juliusson, Per 30, 137
The Jungle Book 30
Jungle Lore 184

Kahani 150–3, 155–7
Kala, Durga 169
Kala Bhairava 13, 39, 40
Kaladhungi 184
Kali 11, 13, 33, 39, 69, 76–80, 103, 148, 162
Kali Yug 1
Kandang Balok 53
Kandhs *see* Khonds
Kapalikas 164
Kapildhara 20
Karanjia 10, 154
Karei 42
Karens 42, 60–1
Kathiawar Peninsula 161
Kedah 42, 113, 125
Kedarnath
Kelantan 37, 52, 113, 129
Kemaman 7, 94, 106
Kensiu 42, 95–6
Kerala 162
Kerinci 28, 38, 42, 43, 46–7, 54, 102, 106, 108, 117–34
Ketari 129–32
Keyser, Arthur 6, 7, 23, 118–19
Khonds 49, 147–50, 158
kiais 118
King Martin, D. 62
King Paroi 53

King Uban 53
Kingscote, Georgiana 111–12
Kinta 132
Kipling, John Lockwood 37, 93, 157, 165–6
Kipling, Rudyard 30
Knebel, Josef 98, 99, 100
Knight, Chris 102
Kochi 162
Kodagu 48
Kodavas 48, **49**
Kola Buloo 56
Kols 10, 31, 67, 139, 140–1, 147
Konkan 76
Koombappa 41
Koran 7, 52, 53, 126
Korea 1, 62
Korkus 10, 12–13, 51, 61, 69
Koyna 76–80
Kuala Kangsar 130
Kuala Langat 40, 52
Kuala Lumpur 39, 127–8, 132
Kuala Terengganu 128
Kuang-i chi 105
kulpa briksha 112, 149–50
Kumaon 167–84
Kushnir, Hadas 103
Kusru Gond clan 33–4, 69, 73
Kutra Pass 158–9

Labu 134
Lahore 138, 166
Lakshmi 112
Lamongan 1, 108
Lao 108
Laos 1, 48
Latham, Robert 30
Laufer, Berthold 31
Leaves from an Indian Jungle 167
Lehman, Frederic K. 48
limes 38, 47, 95, 97
Lindskog, Birger 97
"lion brides" 111, 123
lions 25, 28, 30, 67, 103, 161
Locke, Arthur 7, 29, 32–3, 40, 43, 50, 52, 53, 58, 60, 94, 106, 114, 124, 127, 129, 180
Lodoyo 108, 118
LSD 101
Lumsden Milne, Betty 127, 128
Lurmi Range 81–3, 135–6, 163
lycanthropy 94

macan bumi 30
macan gadongan (gadungan) 102, 158
Madhya Pradesh 9
Mahabaleshwar 76, **77**
Mahadeo 13
Mahadeo Hills 12, 150, 159

204

Index

Mahadevi 33
Maharashtra 76, **77**
mahua spirit 13, 69–71, 73, 162
Maihar 158–9
Maikala Range 11, 20
mail runners 19–20, 53
Malabar 96
Malacca 53, 119
Malcolm, Maj.-Gen. Sir John 139–40
Man 107
Man-Eaters of Kumaon 184
man-eating 3–8, 19–26
Manchuria 35
Mandla 10, 11, 16, 17, 20, 22, 33, 36, 68, 69–74, 87–92, 137, 138, 165
Mannans 162
Marathas 76–80, 154
Markam Gond clan 33
Marsden, William 53, 120
marulupulis 141
Mason, Philip 176
Masters, Robert 101
Matti uthana 70–3, 74, 75
mavis 95, 105
Maxwell, William Edward 124
Maxwell, (William) George 124–6, 127, 128–9
McEldowney, Philip 156
McMillan, Archibald 72–3, 74–5
McNair, Maj. Frederick 23
Mecca 118
Mechas 66
Meghalaya 51
meriah sacrifice 148
Mervyn Smith, A. 144–7
metempsychosis 36
Mhars 79
Mills, James 54
Minangkabau **94**
Minshull, Mrs. E. 160–1
Mitra, Sarat Chandra 33, 34, 35, 43, 50, 51, 52, 165, 166
Mirzapur 69, 158
Mlabri 1, 42, 107, 108
Moguls 77, 154
Mohnike, Otto 53
Moi *see* Muong
moon 62, 73, 78, 80, 84, 87, 97, 102–3, 115, 132, 137, 139, 142, 145, 149
Mother Earth 10; *see also* Earth Mother
Motley, James 120
Mount Angsi 53
Mount Dangka 40
Mount Dempo 53, 120
Mount Kandana 54
Mount Ophir 53
Muduvar 6

Muhammad 28, 37, 42, 47, 54
Muktesar 170
Mulud 37
Mundas 72
Muong 6, 27, 34–5, 42, 47, 50, 51, 53, 56–8 (**57**), 59, 61, 63, 101
Mysore 14, 16, 41, 67, 91, 165

Naga Hills 28, 65–6
Naganijan 141–4
Nagaon 116
Nagas 28–9, 42, 47–8, 49, 54, 95, 100, 103, 104, 105, 108
Nagpur 21, 33
Naini Tal 19–20, 22, 55, 169, 176
naming 101
Narimangala 48, **49**
Narmada 9, 11, 13, 59, 84, 150, 159, 164
Narsinghpur 59
nats 34, 55
Natural History of the Mammalia of India and Ceylon 151
Nauratri 139–40
necklaces 46, 79, 149, 159, 164
Negeri Sembilan 53, 105–6, 113, 119
Negritos 4, 115, 117; *see also* Batek; Kensiu
Nepal 23, 66–7, **68**
Netia Gond clan 33
Neuri 107
Newbold, Thomas 129
Nias 112
Nimar 13, 17
North Sumatra 47, 96
Nou Yu-t'u 31

Odisha *see* Orissa
O'Donnell, Elliott 147–50 (**147**), 158
olthwa 36, 106
Oman, John Campbell 138, 139
Omkar 13
Oraons 142
Orissa 42, 49, 147–50
Osborne & Chappel 132

Pachmarhi 12–14
Packer, Craig 103
Pahang 4, 32, 113–16, 119, 125, 128, 129–32, 134
Palembang 98
Paliyans 52
palmyra 111–12, 150
Panar 170–1, 176
Pandit Natesa Sastri 111–12
Paniyans 96
Pankas 30
panok 95–6
Pante Cermen 94, 98

205

Panti 45
"pard" 151
Partabgarh 76, 79, 80
Pasemah 53, 120
Pashtuns *see* Pathans
Pashupati 5
Pathans 15, 155, 156, 176
Patna 138–9
pawangs 38–40, 46, 47, 60, 119; *see also* shamans
peepul 150
pelsits 36
Penang 119
Perak 113, 116, 119, 121–4, 125, 130, 132–4
Perlis 113
Perry, Richard 35, 59, 116
Petronius 100, 102, 103
pheal 56
pheeow 56
philtrum 98, 107
Phom 54
phuao 56
Pidie 40
Pineapple Hill 132
The Pioneer 177
Piyongkong 54
Pliny the Elder 100
Pootinadi 80–1, 141–4
post-mortem metamorphosis 35–8, 105–6, 109
post runners *see* mail runners
Powell, Arthur 6, 51, 65, 162
Prata 108
Presgrave, Edward 120
priests (Brahman) 5, 41, 76, 79, 159; *see also* Brahmans
priests (tribal) 5, 29, 34, 42, 48, 66–75, 158, 162; *see also* shamans
Pu Pongling 45–6
purveyance system 10–11

rabies 116
Raffles, Lady Sophia 45
Raffles, Sir Stamford 45, 120
Raipur 60
Rajah Yah 54
Rama 112, 149, 150
Ramayana 111
Ramganga National Park 170
rampok macan 118
Rangachari, K. 162
Ranggul 114–16
Rathborne, Ambrose 129
Ratnagiri 76
Reading, Lord **68**
reincarnation 36
Rejang 53, 120

repercussion wounding *see* wound doubling
rewards 22–3
Riau 120
rimau jadi-jadian 102
roots 69–71, 73, 153, 157, 158
Roulet, Jacques 107
The Royal Tiger of Bengal 135
Rudra 167
Rudraprayag 167–84
Russell, Robert 31, 33, 41, 42, 51, 56, 61, 69, 72, 75, 163, 164, 165

sadhus 5, 20, 76, 107, 153, 162–6, 171, 175–7, 183, 184
Sagar 59, 159–60
saliva 95, 148; *see also* spitting
Sanderson, George 25, 41–2, 44, 50, 51, 165
Santhal parganas 143
Santhals 49, 141–4
Sarsan Gond clan 33
Satan 81, 158
Satara 80
Satpuras 9
Savaras 42–3
Sayong 124
Sayyidina Ali 28
Schilling, Ton 98
Scythians 107
Selangor 28, 40, 47, 113, 118, 119
Selene 97
Selim 121–4, 128
Sema Nagas 28–9, 48, 49, 95, 100, 103, 104, 105
Seonee 151
Seoni 41, 150
"shadow soul" 38, 93, 95, 101, 148
Shakespear, Henry 56, 60
shamans 5, 29, 35, 37, 38–40, 46, 52, 54, 66, 93, 95–6, 102, 104, 106, 116–17; *see also pawangs*; priests (Brahman); priests (tribal)
Shans 60–1
Sheikh Mansur 32
Sheppard, Mervyn 113–14
Shiva 5, 11, 13, 28, 33, 38, 55, 76–80, 82, 84, 88, 112, 149, 159, 162, 164, 165
Shivaji 76–80, 154
Shuttleworth, Charles 127, 129, 180
Sichuan 105
silat 93, **94**, 96, 100, 102
silver 106
Singapore 119, 127
Singbaba 30–1
Singhbhum 139
Siputeh 125

sirih 47, 100
Sita 112, 149
Skeat, Walter 28, 40, 47, 52–3, 54, 99, 117, 118, 134, 154
Sleeman, Col. Sir James 170
Sleeman, William 10, 11, 158–60, 162, 170
Society for the Service of the Gonds 10
"somersault tigers" 99, 100, 108
Sonawani Range 21
The Song of Sandsumjee 30
Sonpur 15
sorcerers 69, 96, 107, 135–47, 152–3
South Sumatra 98
spitting 100
The Sporting Rifle and Its Projectiles 14
Sterndale, Robert 30, 150–3, 155–7, 176
Strange Cults and Secret Societies of Modern London 147
Subramani Iyer N. 162
suggestion 101–2
Sukarno 35
Sukhiraj 11
Sunda 36, 37, 40, 54
Sundarbans 34, 166
Surgana 141
Swanson, Alexandra 103
Swettenham, Frank 119, 121, 124, 125, 129

Tai Fu 105
Tamils 111–12
Tangse 97
Tapanuli Selatan 96–7, 105, 120
Tari Pennu 148
Taylor, Mary Linley 62
Tekhu-rho 48
Temple, Sir Richard 17
Terai 34, 66
Terengganu 7, 32–3, 52, 53, 60, 113, 124, 128, 129
Teungku Chik Cicem 40
Teungku Chik di Rakmeh 36–7
Teungku di Krueng 40
Teungku di Kuto 37
Teungku Ulu-n-Tanoh 37
Thailand 1, 39, 42, 55, 107
Thak 171
Tharus 34, 66, 97
Thor bel 70, 73–4, 75
Thugs 11
Thurston, Edgar 6, 42, 49, 52, 96, 141, 148, 153, 162
Thurston, J.B.H. 25, 61, 127, 129
tiger body parts 48–52
tiger charms (spoken) 52–3, 68
tiger clans 30, 31–2, 33–4, 69, 73
tiger deities 33–5, 48, 69, 72–3, 75, 80, 85, 147–50

tiger-form ancestors 35–41, 54, 93, 106; *see also hantu belian*
tiger-form familiars 38–40, 102; *see also* tiger-form ancestors
"tiger magic" 95
tiger numbers 1
Tiger Ravine 98
tiger settlements 53–4, 120–1
tiger skins 5, 49, 99, 101, 105, 108–9, 164
tiger taboos 42–4
"tiger wedding" *see Narimangala*
"tiger's claw" (weapon) **154**
Tigers' Seat **77**
tjindaks (*tjindakus*) 96–7, 105, 106, 107, 109, 110, 120
Tod, Colonel James 44
transmigration 36
traps 23, **24**, 27, 46, 56, 58–9, 65, 78, 91–2, 110, 118, 122, 124–8, 129, 145, 146, 156–7, 169, 172, 174, 175, 179, 180
Travancore 162
Treves, Frederick 184

Untouchables 55, 107
Uribe, Maria 101–2

van Hasselt, Arend 120
van Ossenbruggen, Frederik 112
Varanasi 157
The Varieties of Psychedelic Experience 101
Velans 162
Vietnam 1, **24**, 56
"village tigers" 30, 54
Vishnu 112
The Voice of Jerusalem 27
von Fürer-Haimendorf, Christoph 104

Wardha 116, 154
water 53, 99–100, 120
Wavell, Stewart 39–40, 129–32 (**131**)
Webber, Thomas 19–20, 55
Weld, Frederick 113
weretiger characteristics 98, 105–6
werewolves 95, 98, 100, 102, 103, 105, 106, 107, 110, 150, 158
Werwolves 147, 148
Wessing, Robert 2, 27–8, 29, 32, 35, 36, 37, 38, 40, 42, 44, 46, 47, 51, 52, 54, 56, 63, 80, 81, 94, 95, 96–7, 98, 99, 100, 101, 105, 106, 118, 120, 129, 158
West Java 32, 36, 54, 100
West Sumatra 45
Western Ghats 48, 141
white tigers 76–83
The Wild Sweet Witch 176
Winter, Johannes 99, 100
witches 36, 68, 96, 97, 103, 136–47, 164

Woodruff, Philip 176
wound doubling 103–4, 121–4, 144–7

Yao 108
yogis 153, 164, 165–6

Yogyakarta 35
Young, Gordon 42

Zangwill, Israel 27

www.ingramcontent.com/pod-product-compliance
Ingram Content Group UK Ltd.
Pitfield, Milton Keynes, MK11 3LW, UK
UKHW042001140426
5217IPUK00015B/923